European Consortium for
Mathematics in Industry 1

Hazewinkel/Mattheij/van Groesen (Eds.)

Proceedings of the First European Symposium
on Mathematics in Industry

European Consortium for Mathematics in Industry

Edited by
Michiel Hazewinkel, Amsterdam
Helmut Neunzert, Kaiserslautern
Alan Tayler, Oxford
Hansjörg Wacker, Linz

ECMI Vol. 1

Within Europe a number of academic groups have accepted their responsibility towards European industry and have proposed to found a European Consortium for Mathematics in Industry (ECMI) as an expression of this responsibility.

One of the activities of ECMI is the publication of books, which reflect its general philosophy; the texts of the series will help in promoting the use of mathematics in industry and in educating mathematicians for industry. They will consider different fields of applications, present casestudies, introduce new mathematical concepts in their relation to practical applications. They shall also represent the variety of the European mathematical traditions, for example practical asymptotics and differential equations in Britain, sophisticated numerical analysis from France, powerful computation in Germany, novel discrete mathematics in Holland, elegant real analysis from Italy. They will demonstrate that all these branches of mathematics are applicable to real problems, and industry and universities in any country can clearly benefit from the skills of the complete range of European applied mathematics.

Proceedings of the

First European Symposium on Mathematics in Industry

ESMI I
October 30–November 1, 1985, Amsterdam

Edited by
Michiel Hazewinkel, Amsterdam
Robert M. M. Mattheij, Eindhoven
Embrecht W. C. van Groesen, Enschede

B. G. Teubner Stuttgart

Kluwer Academic Publishers

Editors:

Prof. Dr. Michiel Hazewinkel, CWI Amsterdam
Prof. Dr. Robert M. M. Mattheij, Techn. Univ. Eindhoven
Prof. Dr. Embrecht W. C. van Groesen, Univ. Twente, Enschede

Distributors:

Continental Europe (excluding U.K.)
B.G. Teubner GmbH, P.O. Box 801069, D-7000 Stuttgart-80

United States and Canada
Kluwer Academic Publishers, 101 Philip Drive, Norwell, MA 02061, USA

All other countries (including U.K.)
Kluwer Academic Publishers, P.O. Box 322, NL-3300 AH Dordrecht

CIP-Titelaufnahme der Deutschen Bibliothek

European Symposium on Mathematics in Industry: (October 30 – November 1, 1985 Amsterdam)
Proceedings of the First European Symposium on Mathematics in Industry / ed. by
M. Hazewinkel; R. M. M. Mattheij; E. W. C. van Groesen.
Stuttgart : Teubner
Dordrecht; Boston; Lancaster : Kluwer Acad. Publ. 1988
(European Consortium for Mathematics in Industry; Vol. 1)
ISBN 3-519-02170-6 (Teubner)
ISBN 90-277-2730-9 (Kluwer)

NE: Hazewinkel, Michiel; (Hrsg.) European Consortium for Mathematics in Industry:
European Consortium for ...
Auf der Haupttitels. auch : ESMI ...

Library of Congress Cataloguing in Publication Data

CIP data appear on separate card.

Printed and bound in Germany

Preface

The idea to organise this symposium was induced by the feeling that the gap that exists to some extent between the mathematical professionalists at universities and the practical mathematicians working industry should be closed.

This would be beneficial, both for mathematics as a scientific discipline (by being challenged with hard, but attractive problems from applications), as well as for the understanding of the increasingly complex problems that arise in the technological societies of today.

Triggered by an informal discussion during a workshop at Nijmegen in early 1985, and promising sounds for funding a little later, the aims of the meeting were focussed on three points.

First, on an national level, the meeting could further the existing, mostly informal, contacts between universities and industry, meanwhile investigating the possibilities and necessity to formalize these contracts. A 'Committee for developing industrial mathematics in the Netherlands', chaired by Prof. J.F. Benders, was formed and played an active role to prepare the panel discussion.

The second aim was to directly exchange information with in other, mostly European, countries between University and Industry were being established, in which way university teaching was influenced by such contacts and so on, as well as concerning the kind of problems, and mathematics, that can be expected, by providing typical examples of succesful cooperations.

The actual outcome of this aspect of the meeting is essentially covered by the contents of these proceedings, which contain most of the lectures that were presented.

Recognising the fact that each country on its own has only a limited capacity to tackle interesting problems and to develop special courses in this field, the third aim was to provide a starting point of whatever kind of cooperation on a European scale on these aspects. As a result of this, ECMI (European Consortium for Mathematics in Industry) has been founded in the meantime. More information about this is also contained in these proceedings.

Organising a meeting like ESMI needs financial support. The organising institutes (CWI, Amsterdam and WD, Nijmegen) contributed (of course). The largest amount of support came from the Air Force Office of Scientific Research (London Office) though the good offices of Col. Jack Ferguson, although it was a discussion with Charles Holland of the Office of Naval Research (London Office) that launched the organization. Very substantial support was also received from Nationale Nederlanden and further important support was received from Shell International and, on a contingency basis, the ministry of economic affairs.

To all these we express our thanks. Finally it behoves us to thank the 'Committee of recommendation' (H.L. Beckers, Shell; T.C. Braakman, Nationale Nederlanden; C.J. van Lede, Verbond Nederlandse Ondernemingen; K. Teer, Nat. Lab, Philips), the 'Committee for developing industrial mathematics in the Netherlands', the more general speakers of the scientific-political first day of ESMI: E. van Spieghel, W.A. Koumans, H. Plate and H. Bosma, and, finally, all those who through conferences, panels and discussions contributed to ESMI.

The proceedings contain most of the scientific talks given at ESMI. In addition there are some 40 pages of capsule descriptions of various uses of mathematics in industry. This can be seen as representing, say, the first 1% of a catalogue of what can be done with mathematics. Certainly it is a future task of ECMI to compile such a catalogue. Maybe this ε-sized hopelessly biased beginning will still serve to give a first indication of how varied and how plentiful such applications are.

E. W. C. van Groesen
M. Hazewinkel
R. M. M. Mattheij

Table of Contents

A European Consortium for Mathematics in Industry

Mathematics, as the language of the sciences, has always played a role in technology; the rapidly increasing 'high tech' character of European industry, however, implies that it will in the future become much more important in research and development. 'Too few people recognize that the high technology so celebrated today is essentially a mathematical technology'. This statement by Edward E. David, President of Exxon R & D, is more true now than ever; more and more companies realize that the ability to use powerful computing techniques to solve complicated equations and optimization problems has dramatically changed both organization and simulation of industrial processes and the design of new products. This requires new skills. In the first place mathematical modelling, i.e. the formulation of a 'real world problem' in mathematical terms, and secondly the restructing of this mathematical formulation in a form which will yield qualitative and quantitative answers. This new demand for mathematics has stimulated the founding worldwide of a number of academic groups committed to interaction with industry. Within Europe a number of these groups have accepted their responsibility towards European industry and have proposed to found a European Consortium for Mathematics in Industry (ECMI) as an expression of this responsibility.

The need for a European organisation reflects the repeatedly stated view of industrialists that European industry can only compete in world markets if it can operate on a European scale. It is also a simple matter of fact that the field of applicable mathematics is so large und diverse in knowledge and techniques that no single European country can possibly claim to be able to provide all the expertise needed. The EC also recognizes the need for the exchange of skills and knowledge between member nations. Moreover mathematics is a genuinely international science and the breadth of high level skills required for both research and teaching in mathematical modelling may not be available in a single nation. The major objectives of ECMI have been formulated to respond to these needs of European industry; they may be summarized under three main headings:

A To promote the use of mathematical models in industry.
Large companies usually have several mathematicians on their staff. These can certainly benefit from cooperation with universities by gaining access to the latest information and techniques. As far as medium size and small companies are concerned it is likely that many profitable applications of mathematics are being overlooked for want of appropriate expertise.

B To educate 'industrial mathematicians' to meet the growing demand for such experts.
This education should be available in all member countries at both undergraduate and graduate levels; moreover specially designed 'updating' education of existing industrial staff needs to be developed.

The postgraduate system, especially, will be based on the joint teaching capacity of the ECMI members. It is expressly intended that a substantial part of this programme will be undertaken in another European country other than the student's own.

C To operate on a European scale.
This is to ensure the transfer of knowledge and skills, and effectively to use scarce academic resources in Europe.
Joint European research activities will be organized combining the special knowledge of the ECMI groups. These activities will also promote European industrial cooperation through mathematics.

Details of the services to be provided are appended. Further information about ECMI activities may be obtained from any of the founder members, or from the ECMI centres already established.

After the ESMI meeting in Amsterdam the idea of a European Consortium for Mathematics in Industry (ECMI) took further form at a meeting in Neustadt-Massbach in April 1986 (the "founding meeting"). Meanwhile ECMI also exists as a legal entity (since June 1987). The text above together with appendix 1 below can be seen as a charter document.

A second scientific ECMI meeting took place in Oberwolfach in March 1987 and further meetings are scheduled for 1988 (Scotland), 1989 (Austria), 1990 (Finland).

Appendix: ECMI Services for Industry

1. Problem identification In order to solve a problem through mathematical simulation, one has to be aware of the possibilities offered by mathematics today. Identifying those problems which can be treated mathematically requires scientists with a broad interdisciplinary knowledge, or those who are able to cooperate in an interdisciplinary background knowledge team. ECMI will organize teams of mathematicians whose expertise covers a wide range of both industrial applications and mathematical background: these groups can help companies to use mathematics efficiently. Moreover, normal mathematical education teaches the ability to solve problems posed in mathematical terms; ECMI groups will stress the importance of problem identification.

2. Identification of Expert Advice Efficient problem solving very often requires the use of recent research results in many different mathematical fields; no single mathematician, nor any team, is able to cover the whole domain. However given a problem area, it may only be necessary to know who the relevant expert is. The mathematical community is a very open community, and mathematicians are in general very willing to communicate their knowledge. Having identified a problem, ECMI will know the expert to be asked.

3. Problem Development and Solution ECMI is certainly neither able nor willing to do all mathematical research work for industry. On the other hand, if in different problem areas similar mathematical questions arise whose resolution requires fundamental mathematical research, ECMI will be in a position to establish research groups for this purpose.

4. Training of graduate mathematicians As noted above, there is a growing demand of 'industrial mathematicians'. In response to this demand ECMI is establishing a postgraduate programme "Mathematics for Industry". This programme will provide graduate mathematicians, engineers and natural scientists with a professional application-oriented qualification whose emphasis is on mathematical modelling of technical problems and evaluation of these models. The programme will help them to become "real problem solvers".

5. Further training of industrial staff ECMI is planning the design of special courses on demand from specific industries using available experts. The aim of these courses is to make industrial staff (engineers, natural scientists etc.) acquainted with recent results in applied mathematics and with the capabilities of mathematical modelling of technical problems.

6. Encouragement of students to seek careers in European Industry The European postgraduate programme on Mathematics for Industry will further educate mathematicians and scientists, with little knowledge of applications, into industrial mathematicians. The genuinely European experience which is an essential part of this programme will provide a strong encouragement to seek a European cureer.

On behalf of the ECMI council
M. Hazewinkel
R. Mattheij

ECMI Thoughts

The motivating feelings of the organisors of this conference might be summed up in the following slightly overcharged statement:

"There are very few problems indeed in the worlds of commerce, technology and science where a combination of modelling, mathematical analysis, numerical work and simulation will not yield substantial results. Often at quite moderate cost compared to other ways - if any - of finding out things".

This puts it maybe a shade baldly and ignores e.g. such things as the highly necessary expertise from the problem area involved. Also one may ask what is particularly relevant about the statement with respect to here (Europe) and now (1987). Well, for one thing the insight generating power of mathematical modelling has probably long been underrated by scientists and even more so by technology and commerce. However, at the present time there seem to be at least three additional indicators we can think of. They centre around the three phrases (i) computing power, (ii) tools from pure mathematics, (iii) technological (and other) challenges - increasing mathematization of the sciences and technology. Let us consider these in turn.

Computing power

Having a reasonable mathematical model is a large and nontrivial step in tackling an industrial problem. However this is not enough; it remains to be analyzed, ideally "to be solved". Very few models, however, are exactly (analytically) solvable and even qualitative insights may be very hard to come by. Then computing becomes important in several forms: numerical methods (possibly also using heuristic and interactive models), simulation, "what if ..." trial runs (in decision problems), ..., yielding both precise numerical solutions and (analytical) insight in the more qualitative properties of the model.

It is perhaps appropriate at this stage to recall some words of John von Neumann in 1946.

"The advance of analysis is at this moment stagnant along the entire front of nonlinear problems. ... [This is] not [a] transient [matter] ... we are up against an important conceptual difficulty."

"However, we remark that really efficient high-speed computing devices may in the field of nonlinear partial differential equations as well as in many other fields which are now difficult or entirely denied of access, provide us with heuristic hints which are needed in all parts of mathematics for genuine progress. This should lead ultimately to important analytical advances."

It should be said here that obtaining numerical solutions is definitely not the only use of computing power; it is equally important to experiment numerically to obtain analytic and qualitative results. As R. W. Hamming puts it.

"The purpose of computing is insight, not numbers."

All this of course pertains to nonlinear mathematics. Linear mathematics is powerful and immensely useful, but nonlinear mathematics is where the rewards are. A purely linear world would be a sad place to live in; no computing devices for one thing. The computer, being a powerful experimental mathematical tool has wider repercussions. As a result there are now quantities of new, surprising, often totally unexpected (mathematical) phenomena for mathematicians to think about. Well known examples are universality properties in chaos, and "the soliton revolution". All this works to rapidly increase the number of analytically and/or qualitatively tractable potential models, which in turn will be most useful in industrial mathematics.

Tools from pure mathematics

Let start with a quotation from a preface by Gail S. Young.

"It is close enough to the end of the century to make a guess as to what the Encyclopedia Brittanica article on the history of mathematics will say in 2083.
We have said that the dominating theme of the Nineteenth Century was the development and application of the theory of functions of one variable. At the beginning of the 20th Century, mathematicians turned optimistically to the study of functions of several variables. But wholly unexpected difficulties were met, new phenomena were discovered, and new fields of mathematics sprang up to study and master them. As a result, except where development of methods from earlier centuries continued, there was a recoil from applications. Most of the best mathematicians of the first two-thirds of the century devoted their efforts entirely to pure mathematics. In the last third, however, the powerful methods devised by then for higher dimensional problems were turned onto applications, and the tools of applied mathematics were drastically changed. By the end of the century, the temporary overemphasis on pure mathematics was completely gone and the traditional interconnections between pure mathematics and applications restored."

Thus there are tools, a great variety of powerful tools, whose potency and effectiveness remains to be tested in more applied situations. Many of these "tools" seem esoteric at first sight, e.g. the theory of p-adic fields, a topic in number theory. Yet these nonarchimedian ordenings and absolute values are now finding applications in the study of the group of materials known as spin glasses. Nonstandard analysis is another such esoteric mathematical fancy, but is enormously useful in stochastic optimal control problems (and it yields real results; i.e. results in the standard world). Superspace, where there are both normal, commuting variables and anticommuting variables, is equally weird but is a great help in understanding such a fundamental thing as the Feymann-Kac formula. The question of whether convex regions necessarily contain lattice points originated in number theory (geometry of numbers) and has all sorts of applications to integer programming problems. Cohomology is definitely not a standard applied mathematics technique. Yet it counts numbers of solution of equations, detects obstructions to global stabilizability of nonlinear control systems and helps us keeping track of certain algorithms.

It has often been said that an applied mathematician must, among other things, be a generalist, acquainted with about all of modern mathematics. Clearly impossible, for an individual, but in principle attainable by a multinational consortium. Put in another way, there seems to be no part of pure mathematics immune from being applied and that includes such things as transcendental and irrational numbers (with applications in numerical mathematics).

Technological challenges. Mathematization

Let us again start with a few quotations, this time from the David report.

"Too few people recognize that the "high-technology" that is celebrated today is essentially mathematically technology."

"Industry awareness of the significance of mathematics for technology seems to be increasing. About one fourth of the Ph.D's in the mathematical sciences currently move into industrial careers. The broadly trained mathematician, even at the pre-Ph.D level, is highly employable. Some mathematical research groups in industry are proliferating and the attachment of mathematicians to other groups is growing. As mathematics penetrates into production control and manufacturing through automation, demand for mathematicians will increase; this will place new responsibilities on those who train mathematicians."

This works both ways. The quotations above stress the usefullness of mathematics for designing, managing and controlling industrial processes. On the other hand problems in engineering, physics and chemistry, geology etc. are generating many new mathematical questions requiring new ideas to solve them. Thus for example to solve certain stabilization problems for control systems in industrial plants one would like to have an effective version of the Quillen-Suslin theorem from algebraic K-theory (which says that algebraic vector bundles over an affine space are trivial; never mind what that means). Again stabilizing a linear plant whose parameters may fluctuate a bit brings in techniques from complex variable and interpolation theory. Very practical optimization problems (operations research) have spawned whole new areas of mathematics, much applied, but also involving lots of "pure" questions especially in combinatorial optimization and complexity theory (of algorithms). Flexible manufacturing systems, networks of computers and (infinite) systems of interacting particles all ask probabilistic questions rather different from those asked and answered before and are requiring new techniques and ideas. Geological exploration and image processing are a perfect mine of nice difficult inverse problems and the problems involved in the growth of crystals have been said (J. S. Langer) to have created a whole new body of mathematics.

It is an often observed historic-scientific fact that major discoveries in a science follow soon upon the development of new observation (measuring) instruments. For the first time in its history mathematics now has such a tool and the discoveries are starting to appear. (In Experimental Mathematics, Math. Modelling **6** (1985), 175-211, I have written at length on this).

It is an equally well observed circumstance that the coming together of different strands of thought and tradition is likely to spark of major advances. Here in industrial mathematics, there appear to be several such strands. Thus the future looks very bright indeed, and should be uniquely interesting.

The remarks made above may help to convience reader that now is a good time to put even more efforts in applying mathematics to industrial and technological problems.

We also feel that this has to be done on an European scale, whence the word "European" in the title of the conference. There are several reasons but the main one probably is that no single Western European country (not even one of the mathematical giants) is likely to have all the expertise needed for a broadly based industrial mathematics effort. This applies to "problem solving", to "training students" to "special courses", to "awereness-of-what-is possible courses" for managers and directors, to the creation of a suitable database of experts, to ... . Mathematics is simply a very large field of human endeavour; it is also growing at a rate of some 300000 new theorems a year (not to mention other achievements like algorithms).

Michiel Hazewinkel

APPLICATION OF STOCHASTIC CONTROL TO THE MANAGEMENT OF ELECTRICITY PRODUCTION

A. BENSOUSSAN(*)

Introduction.

The objective of this lecture is to describe a cooperative research between INRIA and EDF, under the responsability of J.P. QUADRAT at INRIA and the Service of Economic Studies at EDF. Under this program, INRIA develops research in stochastic control and complex dynamic systems, with applications to the field of electricity production. The problems of interest for EDF are described and modelled in a collaborative manner, the methodology is mainly developed at INRIA, adapted to the practical problems by EDF, with INRIA help.

Under very broad terms, the major problem of a company like EDF consists in adapting efficiently the electricity production means to satisfy the various needs of customers. The decisions are of three kinds ; short term, medium term, long term. In the long term, one adjusts by appropriate investments the production means to the evolution of the demand. The medium term decisions consist in smoothing the production, given a set of production means, which is not adjustable. In the short term, one adapts the production plan to take into account possible random perturbation.

It is also important to notice that the general trend of production and demand is rigid, even in the long run. However around this general trend there are highly random effects. This explains why regulation is necessary.

1. - A general model of production smoothing.

1.1. Description of the model.

This model has been developed at EDF by TORRION [5]. It is a discrete time model with horizon one year, and period 1 day. Consider N independent regulation means (reserves of water ; price decisions,...), we denote by

(1.1) y_i^t = stock of energy available in the reserve i, at time t, i=1...N

u_i^t = control variable ; production of electricity

a_i^t = random natural inputs of water

d_i^t = energy produced a priori by the reserve i.

(*) University Paris Dauphine and INRIA

We can write the following state equations

(1.2) $$y_i^{t+1} = y_i^t + a_i^t + \rho(u_i^t)^- - (u_i^t)^+ - d_i^t$$

ρ = return of pumping

(1.3) D^t = additional demand of electricity (not supplied by $\sum_i d_i^t$)

The quantity $D^t - \sum_{i=1}^{N} u_i^t$ has to be produced by emergency means, such as available thermal plants).

Call C^t this cost. Hence

$$E \sum_{t=i}^{T} C^t \left(D^t - \sum_{i=1}^{N} u_i^t\right)$$

represents the cost of producing the required electricity, by emergency means.

We can now make explicit the model, formalized as a discrete time stochastic control problem We have

(1.4) $$y_i^{t+1} = y_i^t + a_i^t + \rho\,(u_i^t)^- - (u_i^t)^+ - d_i^t;$$

$$o \leq y_i^t \leq y_i^{max}$$

$$u_i^{min} \leq u_i \leq u_i^{max}$$

$$\text{Min } J = E \;\left\{ \sum_{t=i}^{T} C^t (D^t - \Sigma\;\; u_i^t) + F(y_1^T, \dots y_N^T) \right\}$$

where F represents the value of the stocks of the various reserves available at time T.

This function results generally from a long term optimization problem.

1.2. Methods of solution

Theoretically (1.4) can be solved by the technique of Dynamic Programming. Define

$$V^t\;(x) = V^t\;(x_1 \dots x_N)$$

then the function V^t satisfies the following induction

$$V^t(x) = \underset{D,C,a}{E} \quad \underset{u^t}{\text{Min}} \left\{ C^t \left(D^t - \sum_{i=1}^{N} u_i^t\right) + V^{t+1}\left(x + a^t + \rho\,(u^t)^- - (u^t)^+ - d^t\right)\right\} \tag{1.5}$$

$$V^T(x) = F(x)$$

The serious drawback of D.P. stems from dimensionality. Therefore approximation techniques are necessary. An efficient one used in this context is that of aggregation decomposition. The problem is decomposed into N 2 dimensional problems.

We define

$$\bar{y}_i^t = \sum_{j \neq i} y_j^t \quad , \quad \bar{y}_i^{max} = \sum_{j \neq i} y_j^{max}$$

$$\bar{u}_i^t = \sum_{j \neq i} u_j^t \qquad \bar{u}_i^{max} = \sum_{j \neq i} u_j^{max}$$

$$\bar{u}_i^{min} = \sum_{j \neq i} u_j^{min}$$

we write the state equations

$$y_i^{t+1} = y_i^t + a_i^t + \rho\,(u_i^t)^- - (u_i^t)^+ - d_i^t \tag{1.6}$$

$$\bar{y}_i^{t+1} = \bar{y}_i^t + \bar{a}_i^t + \rho\,(\bar{u}_i^t)^- - (\bar{u}_i^t)^+ - \bar{d}_i^t$$

$$o \le y_i^t \le y_i^{max} \quad , \quad o \le \bar{y}_i^t \le \bar{y}_i^{max}$$

$$u_i^{min} \le u_i^t \le u_i^{max} \quad , \quad \bar{u}_i^{min} \le u_i^t \le \bar{u}_i^{max}$$

with the cost function

$$E\ \{\ \Sigma\ C^t\ (D^t - u_i^t - \bar{u}_i^t) + F(y_i^T, \bar{y}_i^T)\}$$

This problem is solved by D.P. (in 2 dimensions)

$$(1.7) \qquad V_i^t(x_i, \bar{x}_i) = \underset{D, C, a}{E} \ \underset{u_i^t, \bar{u}_i^t}{\text{Min}} \ [\, C(D^t - u_i^t - \bar{u}_i^t) + V_i^{t+1}(x_i^{t+1}, \bar{x}_i^{t+1})\,]$$

where x_i^{t+1}, $\bar{x}_i^{t+1}$ are given by (1.6) in which y_i^t, $\bar{y}_i^t$ are replaced by x_i^t, $\bar{x}_i^t$.

The feedback derived from (1.7) is used for the decision i.

This approximation technique has been very efficient in practice.

2. - Short term management of thermal power plants

2.1. Setting of the model

We describe here a typical short term decision problem. The horizon is one day. Consider a set of thermal stations and hydroelectric stations. We begin with the thermal power stations, whose number is J. Let C_j, $_{j=1..J}$ be the unit production cost of the station j. There is also a start up cost k_j, to be payed just to put the station j in operation (cost of heating time lost for production). Moreover there are working limits $\underline{p}_j$, $\bar{p}_j$.

Assume that the stations are ranked such that $i \geq j \Rightarrow c_i \geq c_j$.

We use the following heuristics : one starts up the power stations in the order of increasin numbers. It results that they are only J+1 possible configurations.

Let us compute the cost of change from state m_1 to state m_2, namely

$$(2.1) \qquad K(m_1,m_2) = \sum_{J=m_1+1}^{m_2} k_j \,, \text{ if } m_2 > m_1$$

$$0 \text{ otherwise,}$$

Next the production cost corresponding to a production p and a state m, denoted by C(m,p), where the range of p is $[\sum_{j=1}^{m} \underline{p}_j \,, \sum_{j=1}^{m} \bar{p}_j]$, is defined by the following expressions :

if m=1,

$$C(1,p) = c_1 p \,, \qquad \underline{p}_1 \leq p \leq \bar{p}_1$$

$$C(o,p) = 0 \quad \Rightarrow \quad p = 0$$

if m ≥ 2, there are two possible cases,

- if there exists h_p such that

$$\sum_{j=1}^{h_p} \bar{p}_j \le p < \sum \bar{p}_j + \underline{p}_{h_p+1} \text{ , then}$$

(2.2) $$C(m,p) = \sum_{j=1}^{h_p} c_j \bar{p}_j + \sum_{j=h_p+1}^{m} c_j \underline{p}_j.$$

- if there exists h_p such that

$$\sum_{j=1}^{h_p} \bar{p}_j + \underline{p}_{h_p+1} \le p < \sum_{j=1}^{h_p+1} \bar{p}_j \text{ , then}$$

(2.2)' $$C(m,p) = \sum_{j=1}^{h_p} c_j \bar{p}_j + c_{h_p+1} (p - \sum_{j=1}^{h_p} \bar{p}_j) + \sum_{j=h_p+2}^{m} c_j \underline{p}_j.$$

Now since the horizon is very short, a continuous time model is preferable. Let

$M(t)$ = state of the system of thermal stations

$M(t) \in \{0 \dots J\}$.

Let $\tau_1, \tau_2, \dots, \tau_l \dots$ the sequence of instants of jumps of $M(t)$. Let also $\mu_1 \dots \mu_l \dots$ the intensities of the corresponding jumps, we can write

(2.3) $$dM/dt = \sum_{\ell} \mu_{\ell} \delta(t-\tau_{\ell})$$

where δ is the Dirac measure. Note the constraints

$$-M(\tau_{\ell}-0) \le \mu_{\ell} \le J-M(\tau_{\ell}-0).$$

Now let

$p(t)$ = production of the thermal stations

$$\sum_{j=1}^{M(t)} \underline{p}_j \leq p(t) \leq \sum_{j=1}^{M(t)} \bar{p}_j \quad \text{if } M(t) \geq 1,$$

$$p(t) = 0 \text{ if } M(t) = 0.$$

We turn now to the description of the reservoirs, whose number is n. Let

$y_1(t)$ = amount of water available in the reservoir i, we have

$$dy_i = g_i(t)dt + \sigma_i(t)dw_i - v_i(t)dt, \tag{2.4}$$

v_i = amount of water turbined

$v_i(t) = 0$ if $y_i(t) \leq \underline{y}_i$.

We denote by $W_i(y_i,v_i,t)$ the amount of electrical energy produced per unit of time by the stations connected to reservoir i. The deficiency of the power network is

$$(D(t) - \Sigma\, w_i\, (y_i,v_i,t) - p(t))^+$$

and the cost to be minimized is given by

$$E[\int_o^T \{C(M(t),p(t)) + \rho[(D(t) - \sum_i w_i - p(t))^+\}\, dt \tag{2.5}$$

$$+\sum_l K(M(\tau_l\text{-}0),M(\tau_l)) + \phi(y(T),T)]$$

2.2. Technique of solution.

The preceding problem is a stochastic control problem, involving continuous controls and impulse controls.

Call $V_m(x,t)$ the infimum of the cost (2.5) over all possible decisions, when the initial state of reservoirs is x and the initial state of thermal power plants is m. By initial state we mean here the state at some arbitrary time $t < T$, the problem being posed between t and T (following the standard invariant embedding technique of D.P.).

Let us define

$$X_m = -\partial V_m/\partial t + \sum_i A_i(t)V_m - H_m(DV_m,x,t)$$

$$Y_m = V_m(x,t) - \inf_{-m \leq \mu \leq J-m} [V_{m+\mu}(x,t) + K(m,m+\mu)]$$

in which the operator $A_i(t)$ is the 2nd order differential operator

$$A_i(t) = - g_i(t)\, \partial/\partial x_i - 1/2\, \partial_i^2(t)\, \partial^2/\partial x_i^2$$

and the non linear operator H_m is given by

$$H_m(q,x,t) = \inf_{v_i,p} \left[-\sum_i q_i\, v_i + C(m,p) + \rho((D(t) - \sum_i w_i(x_i,v_i,t) - p)^+\right]$$

The following complementarity problem obtains

(2.6) $\qquad X_m \leq 0\,,\ Y_m \leq 0,\ X_m Y_m = 0$

with the final condition

(2.7) $\qquad V_m(x,T) = \phi(x,T).$

This problem is called a Q.V.I. (Quasi variational inequality) following A. BENSOUSSAN - J.L. LIONS [1]. In the context of electricity production, the preceding problem has been solved by C. LEGUAY [3].

3. - Optimization of investments in an electrical network.

3.1. Set up of the problem.

The problem considered in this section is a long term decision problem (investment problem). It consists in optimizing the capacities of the lines of the electrical network, to take into account the evolution of the demand. The network is a graph, whose nodes are the regions. The arcs of the graph correspond to the transportation lines. We denote this set by T. We add a dummy sumit 0. Two arcs connect O to a station i, one represents the production and the other the demand. If i is not a production station, there is only one arc connecting O and i (that representing the demand).

We call G the set of production arcs, and D the set of demand arcs.

We then use the following notation, $j \in T \cup D \cup G$, ω = random state,

(3.1) $\qquad g_j(\omega)$ = production level
or flow carried through j
or demand fulfilled

(3.2) $\qquad \check{q}_j(Q_j,\omega)$ = production capacity corresponding to the state ω and an investment in production capacity Q_j

or transportation capacity corresponding to ω and an investment in transportation capacity Q_j

or

$\check{q}_j(\omega)$ = demand

(3.3) $c_j(q_j) =$ production cost of the level q_j
or transportation cost (= 0)
or benefit for selling q_j

$c_j(\overset{\nu}{q}_j - q_j) =$ penalty associated to the demand which is not fulfilled

(3.4) $C_j(Q_j) =$ investment cost for production Q_j

or investment cost in transportation capacity

3.2. Optimization problem.

Let U_j^s, $j \in GT$ be the investment decided in the year s (in production or in transportation), we can write

(3.5) $$Q_j^{s+1} = Q_j^s + U_j^{s+1} \,,\ j \in GT$$

$$U_j^s \geq 0 \qquad j \in GT$$

$$0 \leq Q_j^s \leq Q_j^{\nu}, \qquad j \in T.$$

We then write the network equations

(3.6) $$Aq^s = 0$$

with the constraints

(3.7) $$|q_j^s(\omega)| \leq \overset{\nu}{q}_j^s(Q^j,\omega) \,,\ j \in T$$

(3.8) $$0 \leq q_j^s(\omega) \leq \overset{\nu}{q}_j^s(Q^j,\omega) \,,\ j \in G$$

and A is a matrix made of +1 or -1. The cost function to be minimized is given by

(3.9) $$\underset{\substack{U_j \\ j \in J}}{\text{Min}} \ \sum_{s=1} \beta^s \Big[\sum_{j \in J} C_j(U_j^s) + E \underset{\substack{q_j^s(\omega) \\ j \in GTD}}{\text{Min}} \ \sum_{j \in GD} c_j(q_j^s) \Big]$$

where $J \in GT$ represents a subset of arcs where investments are being considered. The precedi[ng] problem has been solved very efficiently by the **stochastic gradient** method.

4. - Conclusion

The preceding examples show that stochastic control is a relevant theory for dynamic production and investment problems. However in practical problems a serious drawback stems from the size of the state variable. In addition many different techniques can be considered to solve the problem and the user who is neither an expert of stochastic control nor an expert of numerical methods can be discouraged in applying the general methodology hinted in the previous sections. It must be stressed that general packages are not really useful in this context, since one cannot expect them to handle all the diversity which is met in practice. In fact, it is rather towards the techniques of artificial intelligence that the most interesting perspectives arise. At INRIA, J.P. QUADRAT and his team have constructed an expert system to solve stochastic control problems which can be used with little knowledge of both stochastic control and numerical techniques.

REFERENCES

[1] A. BENSOUSSAN - J.L. LIONS, Contrôle impulsionnel et inéquations quasi variationnelles, Dunod, Paris 1982.

[2] P. CHANCELIER, C. GOMEZ, J.P. QUADRAT, A. SULEM, Vers un système expert pour l'identification et l'optimisation de systèmes dynamiques stochastiques, Congrés Inria d'Analyse numérique, Versailles, dec. 85, North-Holland.

[3] C. LEGUAY, Application of stochastic control to the problem of Optimal Energy Management, in Applied Stochastic Control in Econometrics and Management Science, edited by A. Bensoussan, P. Kleindorfer, C.S. Tapiero, North-Holland, 1980.

[4] DODU, GOURSAT, HERTZ, QUADRAT, VIOT, Méthodes de gradient stochastique pour l'optimisation des investissements dans un réseau electrique, EDF bulletin de la DER, série C, Math. Infor. N° 2, 1981, p. 133-164.

[5] TORRION, Comparaison de méthodes d'optimisation pour la gestion du système de production électrique français, note EDF, 1985.

APPLICATION OF MATHEMATICAL MODELS IN PHARMACEUTICAL INDUSTRY : IDENTIFICATION AND OPTIMAL CONTROL APPLIED TO THERAPEUTICS.

Y. CHERRUAULT, I. KARPOUZAS
MEDIMAT - Université Paris VI
15, rue de l'Ecole de Médecine
75270 PARIS Cédex 06

ABSTRACT. This paper deals with optimal therapeutics defined from mathematical models in pharmaceutical industry. We first make an identification and a optimal control, of linear and nonlinear compartmental models and the other models associated to a functional relation. Then we apply the theoritical obtained results to different models. An original optimization technique giving the global optimum will be described.

RESUME. Dans cet article nous nous proposons d'étudier diverses méthodes pour définir des thérapeutiques optimales à partir de modèles mathématiques de l'industrie pharmaceutique. Nous faisons d'abord une identification et un contrôle optimal des modèles compartimentaux linéaires ou non linéaires et aussi pour des modèles associés à une relation fonctionnelle. Puis, nous appliquons les résultats théoriques obtenus à des médicaments différents. En particulier, nous décrirons une méthode d'optimisation globale.

1. IDENTIFICATION IN LINEAR COMPARTMENTAL MODELLING

Let us consider the following general n-compartments model :

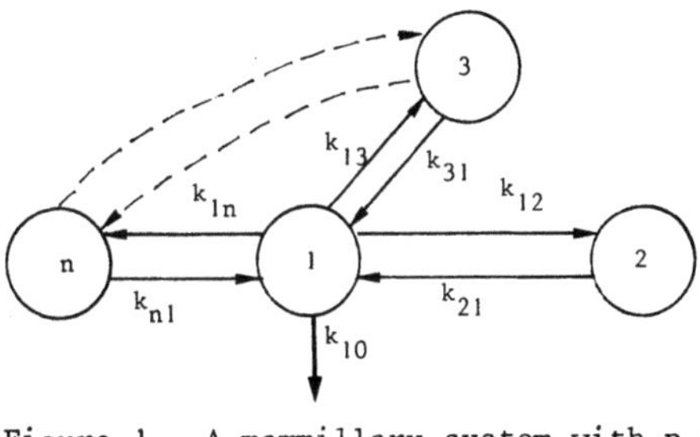

Figure 1. A mammillary system with n compartments.

We know that [10] the structure of this system is translated by a set of n linear differential equations as follows :

$$\left.\begin{aligned} &\frac{dx_1(t)}{dt} = -\sum_{i=1}^{n} k_{1i}x_1 + \sum_{i=2}^{n} k_{i1}\, x_i \\ &\frac{dx_i(t)}{dt} = k_{1i}\, x_1 - k_{i1}\, x_i , i = 2,3,..,n \; k_{11}=k_{10} \\ &\text{with } x_i(0) = 0 \; , \; i = 2,3,.., \; n \end{aligned}\right\} \quad (1.1)$$

Our first problem consists to identify the speed constants k_{ij}, from a partial knowledge about the system in the form of given concentrations $x_1(t)$ in the first compartment (1).

We can write the general solution of the system (1.1) as follows :

$$x_i(t) = \sum_{k=1}^{n} \beta_k^i \, e^{\lambda_k t} \quad (1.2)$$

$$i = 1 \text{ to } n \; , \; \beta_k < 0 \; , \; k = 1 \text{ to } n$$

The identification problem (i.e. the determination of the k_{ij}'s) can be realized by steps.

For identifying the unknown β_k^1 and λ_k we introduce the functional :

$$Y = \sum_{j=1}^{N} (x_1(t_j) - \sum_{k=1}^{n} \beta_k^1 \; e^{\lambda_k t_j})^2 \quad (1.3)$$

$$z_0 = \sum_{j=1}^{N} x(t_i)$$

$$\delta = \frac{1}{z_0}\sqrt{YxN}$$

which has to be globally minimized according to the β_k^1's and λ_k's.

N is the number of observations made for the drug concentrations $x_1(t)$ in compartment 1. Our technique involves a transformation of the function $Y(\lambda_1,...,\lambda_n, \beta_1^1,..., \beta_n^1)$ into the function $y(\theta,\beta_1^1,..., \beta_n^1)$, $\theta \geqslant 0$ using a transformation (called Alienor) [5], reducing n variables to a single one. The absolute minimum of δ will be determined by seeking the best θ minimizing $y(\theta,\beta_k^1)$ where for θ fixed the β_k^1 are calculated by the linear algebraic system $\frac{\partial y_1}{\partial \beta_k} = 0$, k = 1,..., n.

The details of the Alienor transformation [5] will be given in the following.

Then (1.1) and (1.3) involves the following algebraic system :

$$\lambda_r \beta_r^i = \sum_{j=1}^{n} k_{ji} \beta_r^j, \quad i = 1 \text{ to } n, \quad r = 1 \text{ to } n \tag{1.4}$$

Introduce a functional $J(k_{ji}, \beta_r^j)$ such that :

$$J(k_{ji}, \beta_r^j) = \sum_i \sum_r (\lambda_r \beta_r^i - \sum_j k_{ji} \beta_r^i - \sum_j k_{ji} \beta_r^j)^2 + $$

$$+ \sum_{i=2}^{n} (\sum_{j=1}^{n} \beta_j^i)^2 \tag{1.5}$$

Keeping k_{ji} fixed, we minimize J in relation with β_r^j by writing $\frac{\partial J}{\partial \beta_r^j} = 0$ for all r and j. We obtain a system of linear relations giving the β_r^j's. In the second step, we minimize J in relation to k_{ji} by using the relations $\frac{\partial J}{\partial k_{ij}} = 0$ and this process is repeated until we obtain a convergence in the successive values of J. For the initial values of the k_{ij}'s, we have to satisfy the following constraints :

$$k_{ij} > 0 \text{ and } \sum_{i=1}^{n} k_{1i} + \sum_{i=2}^{n} k_{i1} = - \sum_{k=1}^{n} \lambda_k .$$

REMARK. We can consider more general systems such that the following (Fig.2) where we introduce a lag time between the 2 compartments [27] (Enterohepatic circulation).

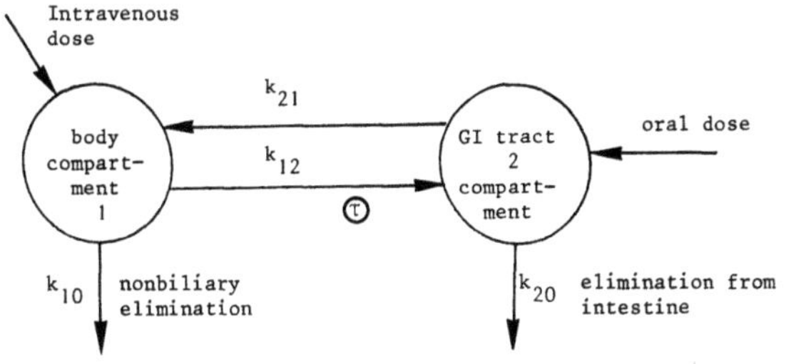

Figure 2. A two compartmental open model with a lag time between the 2 compartments.

The associated model is given by the relationships :

$$\left.\begin{aligned} \frac{dx_1}{dt} &= -\ (k_{10}+k_{12})x_1(t)+k_{21}x_2(t) \\ \frac{dx_2}{dt} &= -\ (k_{20}+k_{21})x_2(t)+k_{12}x_1(t-\tau)\cdot\ Y(t-\tau) \end{aligned}\right\} \quad (1.6)$$

Due to the existence of this time delay the initial conditions take a special form. For intravenous administration :

$x_1(0) = D$

$x_1(t) = 0$ for $-\ \tau \leqslant t < 0$

$x_2(0) = 0$

and for oral administration :

$x_1(t) = 0$ for $-\ \tau \leqslant t \leqslant 0$

$x_2(0) = D$

$Y(t)$ is the Heaviside function (= 0 if $t < 0$, = 1 if $t \geqslant 0$).

2. IDENTIFICATION OF NONLINEAR COMPARTMENTAL MODELS

Consider for instance the nonlinear 2-compartments model represented by the general differential system [13], [16] :

$$\left.\begin{aligned} &\dot{x}_1 = f\ (x_1,\ x_2) \\ &\dot{x}_2 = g\ (x_1,\ x_2) \\ &f(0) = \gamma, g(0) = \delta\ ,\ \text{with}\ \gamma,\delta\ \ \text{constants.} \end{aligned}\right\} \quad (2.1)$$

where f and g are nonlinear functions and with $x_1(t)$ measured for $t \geqslant 0$.

REMARK. If $x_1(t)$ was only measured at $t=t_j$, $j = 1,\dots,m$ then the previous method for identifying the β_k^1 and λ_k could be adapted at this case.

First we choose a mathematical structure for f and g (polynomial functions for instance) and thus the identification problem consists to determine the parameters of the structure. By the Alienor transformation we can reduce the unknown parameters to a single one say θ.

A numerical identification method can be set. It consist to minimize :

$$\int_0^\infty (x_1(t) - x_1^c(t))^2 \, dt \qquad (2.2)$$

according to θ.

In (2.2) $x_1^c(t)$ represents the $x_1(t)$ calculated from (2.1) when the k_{ij} (or equivatently θ) are given. Of course $x_1^c(t)$ depends implicitly of θ.

3. OTHER MODELS ASSOCIATED TO A FUNCTIONAL RELATION

For mathematical modelling it is often useful to introduce a mathematical relationship between the input and the output of the studied system [29]. For instance in pharmacokinetics one needs to find a mathematical relation between the blood concentration and the effect of a drug [9].

Let us consider the following diagram where the exchanges may be nonlinear.

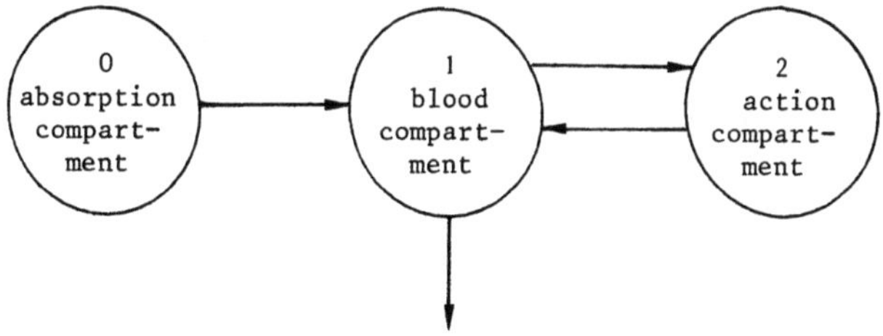

Figure 3. Schematic representation of a three compartment open model.

For determining a nonlinear relationship associated to a concrete β-blocker (studied by SERVIER laboratories), we proposed the following formula relating the dose D, the plasmatic concentration C and the effect E [8] :

$$E = b_o + b_1 \text{ Log } (1 + C(\gamma_D(t))) \qquad (3.1)$$

where b_o is a constant depending on individual, b_1 has to be identified as well as the function $\gamma_D(t)$ that plays the role of a delay function and which depends on the absorbed dose D. In (3.1) only $\gamma_D(t)$ is function of D. More precisely, the two following optimization problems have to be solved [21] :

$$1)\ \min_{\alpha_1,\alpha_2,\lambda_1,\lambda_2} \sum_{j=1}^{N} \left(\sum_{i=1}^{2} \alpha_i e^{-\lambda_i t_j} - C(t_j) \right)^2$$

(approximation of C(t) by a linear combination of exponentials).

$$2)\ \min_{b_1,\alpha_1,\ldots,\alpha_m} \sum_{j=1}^{N} (b_o+b_1 \operatorname{Log} (1+C(\gamma_D(t_j)))- E (t_j,D))^2$$

for each dose D and with $\gamma_D(t) = \sum_{k=1}^{m} \alpha_k(D)t^k$ (3.2)

where m=1,2 or 3. Then the function $\alpha_k(D)$ in (3.2) are identified by using once more an optimization technique [23].

4. ALIENOR OR A GLOBAL MINIMIZATION METHOD

As said previously we suggest a transformation for solving global optimization problems. We know that many systems in biology or medicine lead to identification or optimization problems.

Their resolution needs to use algorithms of minimization. Most of them [32] give only local minimum and need the existence of some derivatives. The technique described below may be applied to continuous functions. It is based on a space filling curve technique. More precisely consider a two variables function to minimize :

$$\min_{x,y} f(x,y) \tag{4.1}$$

Setting

$$x = a\theta\cos\theta,\ y = a\theta\sin\theta,\ \theta \geqslant 0 \text{ (Alienor transformation)} \tag{4.2}$$

we obtain a new optimization problem according to θ .

$$\min_{\theta} f(a\theta\cos\theta, a\theta\sin\theta) = \min G(\theta) \tag{4.3}$$

where $G(\theta) = f(a\theta\cos\theta, a\theta\sin\theta)$.

More generally it would be possible to correct r and θ (polar coordinates) par a general formula $r = h(\theta)$.

It is easy to prove [3] that the optimum θ^* leads to $x^* = a\theta^*\cos\theta^*$, $y = a\theta^*\sin\theta^*$ converging towards the set (x^+,y^+) realizing the minimum of f(x,y). Of course θ may be deduced from (x,y) by the formula :

$$\theta = (1/a)\sqrt{x^2+y^2} \tag{4.4}$$

This method is based on the properties of the Archimed's spiral $\tau = a\theta$ giving an approximation of the Peano's curve when $a \to 0$.

The generalization to n variables can be done without difficulty. A tree structure is obtained [6] leading to the minimization of a single variable function. This method can also give all the maximums or minimums of a n-variables function. Moreover it may be used for approximating n-variables functions and consequently for solving partial differential equations. For example Alienor was applied to a partial differential system coming from a biological system [14] (gas exchanges during human breathing).

When identifying compartmental models [3], the following functional is considered :

$$J = \sum_{j=1}^{m} \left[x(t_j) - \sum_{i=1}^{n} \alpha_i e^{-\lambda_i t_j} \right]^2 \tag{4.5}$$

$x(t_j)$ given, and J minimized according to the α_i and λ_i. Alienor technique consists to bring back the nonlinear parameters λ_i to a single parameters θ, [20]. Numerical programs (in Basic), are available at MEDIMAT laboratory. We shall see many other applications in the following devoted to nonlinear optimization problems.

REMARK. The Alienor transformation (4.2) may be used for solving partial differential equations and more generally functional equations depending at least on two variables.

Our previous transformation leads to :

$$\left.\begin{aligned} x &= a\theta\cos\theta \\ t &= a\theta\sin\theta \\ \theta &\geqslant 0 \end{aligned}\right\} \Rightarrow \theta = \frac{1}{a} \sqrt{(x^2+t^2)}$$

for two variables.

For three variables x,y,z, we have :

$$x = a\theta_1\cos\theta_1$$
$$y = a\theta_1\sin\theta_1$$
$$\theta_1 = a\theta\cos\theta$$
$$z = a\theta\sin\theta \ , \ \theta \geqslant 0$$

and conversely we obtain

$$\theta = \frac{1}{a} \sqrt{z^2 + \frac{1}{a^2} (x^2+y^2)}$$

This transformation can be easily generalized to n-variables $(n \geqslant 2)$, and applied for approximating a general function of n-variables: $g\ (x_1,\ldots,x_n)$. To do that we express $x_1,\ldots,x_n$ in function of θ and thus g becomes a function $G(\theta)$ of the single variable θ . This last function G(θ) can be approximated by [29] :

a polynomial function
a logarithmic development
an exponential development ... or other convenient approximations well adapted to the functions of one variable.

If we want to solve a functional equation Au = f (with u = u(x,t), f = f(x,t)) with some initial A_1 and boundary condition A_2 the following penalty function can be introduced [11], [12], [24] :

$$J_\varepsilon = ||A(U(\theta)) - F(\theta)||^2_{L^2(\Omega)} + \frac{1}{\varepsilon_1} ||A_1||^2_{L^2(\Omega_1)} + \frac{1}{\varepsilon_2} ||A_2||^2_{L^2(\Omega_2)}$$

and minimized according to the α_i : $\underset{\alpha_i}{\text{Min}}\ J_\varepsilon$,

$$\text{with} \begin{cases} \varepsilon = (\varepsilon_1, \varepsilon_2) \to 0 \\ \Omega = \{ \theta \ / \ AU(\theta) = F(\theta) \} \\ \Omega_1 = \{\theta |\ U(\theta) \text{ satisfies } A_1\} \\ \Omega_2 = \{\theta |\ U(\theta) \text{ satisfies } A_2\} \end{cases}$$

and where the α_i are associated to an approximation of U(θ) such that : $U(\theta) = \alpha_1\ \theta + \ldots + \alpha_n\ \theta^n$. We recall that U(θ) correspond to an approximation of u(x,t) when the Alienor transformation is used for reducing (x,t) to θ. A discretization technique leads to :

$$\underset{\alpha_i}{\text{Min}} \{ \sum_j [A(\frac{\alpha_1}{a}(x_j^2+t_j^2)^{1/2}+\ldots+\frac{\alpha_n}{a^n}(x_j^2+t_j^2)^{n/2}) - f(x_j,t_j)]^2 + \frac{1}{\varepsilon_1} \sum_k A^2_{1,k} + \frac{1}{\varepsilon_2} \sum_m A^2_{2,m} \}$$

where x_j denotes $x(t_j)$.

If the operator A is linear, the functional problem is transformed into the linear algebraic system :

$$\frac{\partial J}{\partial \alpha_i} = 0 \ , \ i = 1,\ldots,n \ [7].$$

Otherwise, we can find the absolute minimum of the function J using once more the global optimization method. This method was particularly tested on the heat equation [19] :

$$\frac{\partial u}{\partial t} = u_o \frac{\partial^2 u}{\partial x^2} \quad \text{with } u_o = 1$$

and with the boundary conditions :

$$u(0,t) = u(\pi,t) = 0$$

and the initial conditions: $u(x,0) = 2\sin x,\ 0 \leqslant x \leqslant M$.

For the heat equation, with boundary and initial conditions, the penalty technique with n=12, a step of space equal to 1/10 and a step of time equal to 0,05 has given small relative errors (10^{-3} in mean with some peaks at 10^{-2}).

5. OPTIMAL CONTROL OF LINEAR COMPARTMENTAL MODELS

Consider a drug satisfying a linear compartmental model such as the following (Fig.4) [34] :

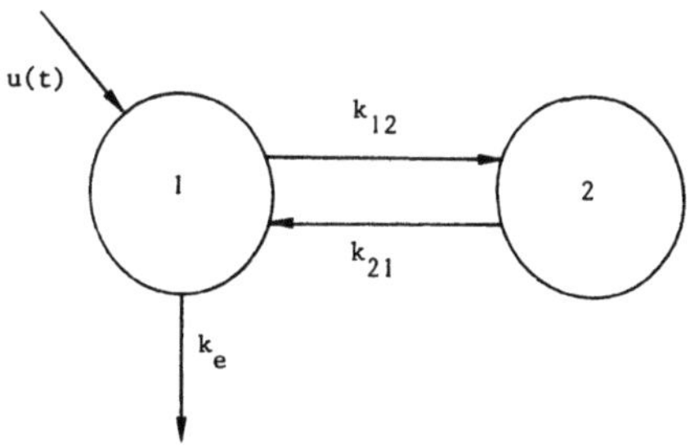

Figure 4. Schematic representation of a two compartment open model. u(t) is an input function (control variable).

where the first compartment is the blood compartment and the second may be the action compartment.

The activity of the drug, being dependent upon it's chemical form, could decrease in kinetically identical ways by transport to another spatial region (elimination) or by transformation to another chemical form (inactivation). Thus compartment is here defined as a state characterized by spatial localization and (or) chemical nature.

From (Fig.4) we have the following system :

$$\left.\begin{array}{ll} \dot{x}_1 = -(k_{12}+k_e)\, x_1 + k_{21}\, x_2 + u(t) & \text{(a)} \\ \dot{x}_2 = k_{12}x_1 - k_{21}x_2 & \text{(b)} \\ x_1(0) = x_2(0) = 0 \text{ (c) fixed initial conditions} & \end{array}\right\} \quad (5.1)$$

where x_i is the amount of the substance (number of particles) in the ith compartment, the constant k_{ji} is the fraction of the susbtance in the jth compartment transported into the ith compartment per unit time, and u(t) is an input function (control variable) which can be controled and chosen so that a given criterion be optimized.

Our problem is the following : Maintain a given quantity A (fixed) in the first compartment by choosing a convenient function u(t) i,e. :

$$\left.\begin{array}{l} \text{Find } u(t) \text{ such that :} \\ \displaystyle\int_0^\infty (x_1(t) - A)^2\, dt \ \text{ be minimum according to u and with constraints (5.1)} \end{array}\right\} \quad (5.2)$$

in a mathematical formulation.

In the case of I.V injection (5.2) possesses an explicit solution [4] given by :

$$u(t) = A\,(\delta_{(0)} + k_e + k_{12}\, e^{-k_{21}t})$$

where $\delta_{(0)}$ is the Dirac mass [28] at time 0 corresponding to the initial conditions $x_1(0) = A$, $x_2(0) = 0$.

Indeed, the second equation of (5.1) with initial conditions $x_1(0) = A$ and $x_2(0) = 0$ may be integrated and gives :

$$x_2(t) = -A(k_{12}/k_{21})\,(e^{-k_{21}t} - 1)$$

Then the first equation of (5.1) leads to the optimal solution for u(t) :

$$u(t) = A\,(k_e + k_{12}\, e^{-k_{21}t})$$

More generally if we consider the general initial conditions : $x_1(0) = x_2(0) = 0$, the Laplace transformation [28] allows to find a general solution which is a distribution function : $u(t) = A\delta_{(0)} + u_1(t)$, where $u_1(t)$ is a classical function solution of (5.1), (5.2) with initial conditions $x_1(0) = A$, $x_2(0) = 0$. Therefore the general solution of our control problem is a distribution function given by :

$$u(t) = A\,(\delta_{(0)} + k_e + k_{12}\, e^{-k_{21}t}).$$

The distribution term correspond to a quantity A put at time 0 in the first compartment. In the case of an oral dose (Fig.5).

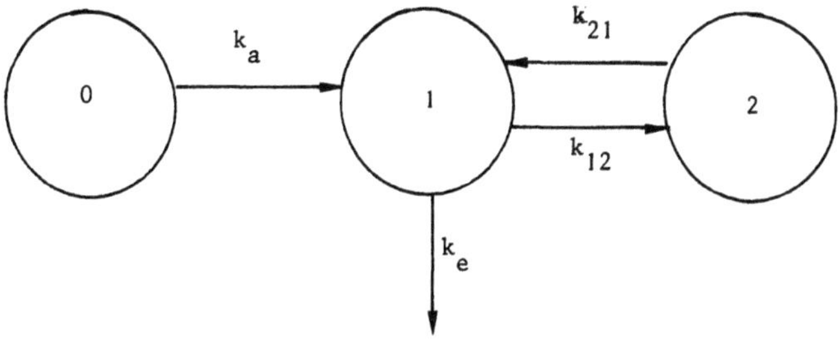

Figure 5. Schematic representation of a three compartment open model with first order absorption k_a where the elimination occurs from compartment (1).

we can associate the following optimization problem :

$$\left.\begin{array}{l} \text{Find } D_i,\ t_i \text{ such that :} \\ \displaystyle\int_0^{\infty} (x_1(t)-A)^2\, dt \quad \text{be minimum, where A is a fixed constant and where } x_1(t) \text{ is the concentration in the first compartment (1).} \end{array}\right\} \qquad (5.3)$$

Problem (5.3) can be solved numerically if we precise $x_1(t)$ corresponding to multiple dose D_j given at t_j . For instance we can write :

$$\sum_{j=0}^{m} D_j\, k_a\, e^{-k_a(t-t_j)} \, . \, Y\,(t-t_j) \qquad (5.4)$$

In (5.4) we suppose that when successive choses are absorbed the corresponding concentrations of the drug are added. The term is thus the resulting concentration, when (m+1) doses D_j are absorbed at time t_j.

One can also try to find D_i and t_i such that :

$$\left.\begin{array}{l} J = \displaystyle\int_0^{\infty \text{ ouT}} \Big(\sum_{j=0}^{m} D_j\, k_a\, e^{-k_a(t-t_j)} \,. \\ \quad .\ Y(t-t_j) - Ak_e - Ak_{12} e^{-k_{21}t} \Big)^2\, dt \end{array}\right\} \qquad (5.5)$$

In (5.5) the term $D_j k_a e^{-k_a(t-t_j)}$ corresponds to an oral dose D_j taken at time t_j, according to the model (Fig.5). Y(t) is the Heaviside function (=0 if $t < 0$, = 1 if $t \geq 0$). Practically (5.5) is equivalent to the problem of finding the best doses and times which give the best approximation of the exact optimal solution

$$Ak_e + Ak_{12} e^{-k_{21}t} .$$

6. CONTROL OF NONLINEAR COMPARTMENTAL MODELS

Consider the nonlinear compartmental model (n-compartments) described by the differential system [13], [25] :

$$\left.\begin{aligned} \dot{x}_1 &= f_1(x_1,\ldots,x_n) + u(t) \\ \dot{x}_2 &= f_2(x_1,\ldots,x_n) \\ &\vdots \\ \dot{x}_n &= f_n(x_1,\ldots,x_n) \\ x_1(0) &= A \ , \ x_2(0) = \ldots = x_n(0) = 0 \end{aligned}\right\} \qquad (6.1)$$

where the f_i's are nonlinear functions.

The associated control problem could be :

$$\underset{u(t)}{\text{Min}} \int_0^{\infty} (x_1(t) - A)^2 \, dt \qquad (6.2)$$

An explicit solution can be easily obtained. Indeed putting $x_1(t) \equiv A$ in (6.1) allows to solve the (n-1) last differential equations according to $x_2,\ldots,x_n$. Let us note $x_2^*,\ldots,x_n^*$ the solution. Then the optimal control u(t) is given explicitly by the first equation of (6.1), i.e. :

$$u^*(t) = - f_1(A,x_2^*,\ldots, x_n^*). \qquad (6.3)$$

The functions $x_2^*(t),\ldots, x_n^*(t)$ being previously calculated (numerically). If we set $x_1(0) = \ldots = x_n(0) = 0$ then the term $A\delta_{(0)}$ has to be added to the optimal solution (6.3).

For instance let us introduce the following two compartments model with Michaelis-Menten elimination kinetics (Fig.6) [1], [16], [33] :

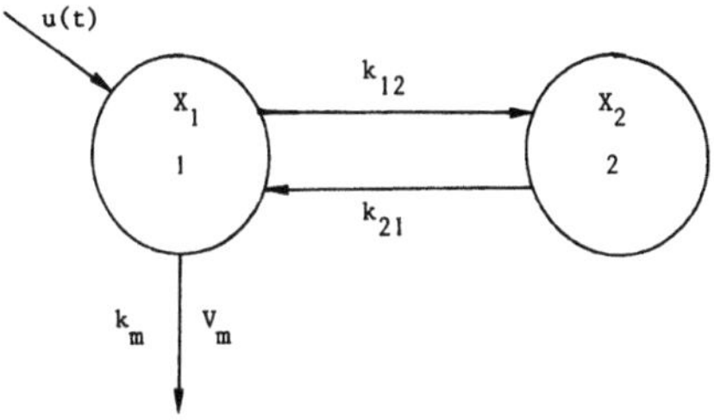

Figure 6. A two compartment open model with Michaelis-Menten elimination kinetics.

The corresponding differential equations are :

$$\left.\begin{aligned} \dot{x}_1 &= g(x_1) + k_{21}x_2 + u(t) \\ \dot{x}_2 &= k_{12}x_1 - k_{21}x_2 \\ x_1(0) &= B \ , \ x_2(0) = 0 \end{aligned}\right\} \qquad (6.4)$$

with $g(x_1) = -(k_{12} + \frac{V_m}{k_m+x_1})\ x_1$ a nonlinear function

The associated control problem will be :

$$\underset{u(t)}{\text{Min}} \int_0^\infty (x_1(t) - B)^2\ dt \qquad (6.5)$$

B is a fixed constant determined by biologists.

For $x_1(t) \equiv B$, $t > 0$ we take from the second equation of (6.4) :

$$x_2(t) = -\ B\ (k_{12}/k_{21})\ (e^{-k_{21}t} - 1) \qquad (6.6)$$

and thus the optimal solution u(t) of (6.4) and (6.5) is given by :

$$u(t) = -\ g(B) + B\ k_{12}\ (e^{-k_{21}t} - 1) \qquad (6.7)$$

Another method would be possible by using a linearization technique [15], [17].

Indeed a Taylor's development of order 1 applied to the system (6.4) gives :

$$\left.\begin{aligned}
\dot{x}_1 &= g(x_1(0))+(x_1(t)-x_1(0))\ g'(x_1(0))+k_{21}x_2+u(t) \\
\dot{x}_2 &= k_{12}x_1 - k_{21}x_2 \\
x_1(0) &= B\ ,\ x_2(0) = 0
\end{aligned}\right\} \tag{6.8}$$

on the interval [0,h] with h (discretization step) small. As previously $x_1 \equiv B$ for $t \in [0,h]$ gives $x_2(t)$ by the same formula

$$x_2 = -\ B(k_{12}/k_{21})\ [e^{-k_{21}t}-1]$$

and the first equation of (6.8) involves :

$$u(t) = -\ g(B) - k_{21}x_2\ . \tag{6.9}$$

On the second interval [h,2h] a Taylor's development of order 1 leads to the linear differential system :

$$\left.\begin{aligned}
\dot{x}_1 &= g(x_1(h))+(x_1(t)-x_1(h))\ g'(x_1(h))+k_{21}x_2(t)+u(t) \\
\dot{x}_2 &= k_{12}x_1 - k_{21}x_2 \\
x_1(h) &= B\ ,\ x_2(0) = 0
\end{aligned}\right\} \tag{6.10}$$

We take the same optimal solution u(t) given by (6.9).

7. CONTROL OF NONLINEAR COMPARTMENTAL MODELS USING A NEW GLOBAL OPTIMIZATION TECHNIQUE

Consider the following nonlinear compartmental model :

$$\left.\begin{aligned}
\dot{x}_1 &= f_1(x_1,\dots,\ x_n,\ u,t) \\
&\ \vdots \\
\dot{x}_n &= f_n(x_1,\dots,\ x_n,\ u,t) \\
&\text{with initial conditions} \\
x_1(0) &= \alpha_i\ ,\ i = 1,\dots,\ n\ (\alpha_i \text{ known constants})
\end{aligned}\right\} \tag{7.1}$$

and the associated optimal control problem :

$$\min_{u(t)} \int_0^T g(x_1(t)\ ,\dots,\ x_n(t),\ u,t)\ dt\ ,\ g \text{ given} \tag{7.2}$$

where the vector $x(t) = (x_1(t),\dots,\ x_n(t))$ and the function u(t) satisfy the constraints (7.1)

The model $(f_n, \ldots, f_n)$ is supposed previously identified (known)

First the Alienor transformation is applied for connecting the variables $x_1, \ldots, x_n$. These variables are transformed into a single variable $\theta(t)$.

Now we use the relations :

$$\underset{u}{\mathrm{Min}} \int_0^T g(x_1(t), \ldots, x_n(t), u, t)\, dt =$$

$$\underset{u_1, \ldots, u_n}{\mathrm{Min}} \sum_{i=1}^{n} \int_{t_i}^{t_{i+1}} g(x_1(t), \ldots, x_n(t), u_i, t)\, dt$$

$$\# \underset{u_1, \ldots, u_n}{\mathrm{Min}} \sum_{i=1}^{n} (t_{i+1} - t_i)\, g(x_i, u_i, t_i) =$$

$$\sum_{i=1}^{n} (t_{i+1} - t_i) \underset{u_i}{\mathrm{Min}}\ g(x_i, u_i, t_i) \qquad (7.3)$$

obtained from a numerical integration formula.

From (7.2), (7.3) and with the help of Alienor transformation, we obtain the following minimization problem :

$$\underset{u_i}{\mathrm{Min}}\ g(P_1(\theta), \ldots, P_n(\theta), u_i, t_i) \qquad (7.4)$$

Solving (7.4) leads to the identification of the function :

$$u_i^* = u_i^* (\theta, t_i) \qquad (7.5)$$

where u_i^* is solution of (7.4), $i = 1, \ldots, n$

In fact u_i is a function independent of i because we minimize g independent of i.

If $\theta = z\ (x_1, x_2, \ldots, x_n)$ the inverse Alienor transformation we obtain :

$$u_i^* = u_i^* (x_1, \ldots, x_n, t_i)$$

which is carried back in the system (7.1). Then (7.1) depends only on $x_1, \ldots, x_n, t$ and it's numerical resolution gives the optimal state functions $x_1^*, \ldots, x_n^*$. Putting these values into (7.5) leads to the optimal solution $u^* = u^*(\theta^*, t_i)$, where θ^* is calculated easily from $x_1^*, \ldots, x_n^*$.

REMARK. A more simple method consists to solve :

$$\operatorname*{Min}_{u} \; g(x_1, \ldots, x_n, u, t) \tag{7.6}$$

by using the Alienor transformation reducing $(x_1,\ldots,x_n,t)$ to θ.

g becomes $G(\theta,u)$ and (7.6) is transformed into :

$$\operatorname*{Min}_{u} \; G(\theta,u) \tag{7.7}$$

leading to $u^*(\theta)$ such that :

$$G(\theta, u^*(\theta)) = \operatorname*{Min}_{u} \; G(\theta,u)$$

It is obvious that $u^*(\theta)$ minimizes the criterion of (7.2).

Then the previous method can be applied for the numerical calculus of $x_1^*,\ldots,x_n^*$. We obtain the optimal control $u^*(\theta^*)$ with the same technique than previously. We can easily generalize to a nonlinear differential system where u is a vector instead of a scalar.

8. NUMERICAL RESULTS

8.1. Enterohepatic circulation (Dr FLOUVAT, Hôpital Ambroise Paré, Paris)

When a substance excreted into the bile passes through the lumen of the intestine, is reabsorbed and then is carried to the liver via the blood flow, it undergoes enterohepatic circulation or cycling. Many endogenous and exogenous substance (bile, salts, morphine, methadone, digitoxin) can undergo enterohepatic circulation [2], [26].

Results of our simulation are drown in Fig.7. Lag time found : $\tau = 7.92$ (h. h/100).

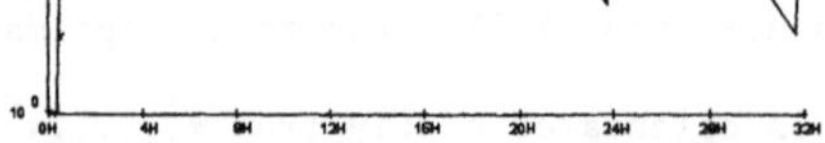

Figure 7. Simulation of the concentration x_1 of the drug. The vertical axis representents the concentration of the drug in mg/ℓ. The horizontal axis representents the time in hours.

8.2. The β-blockers action (SERVIER laboratories - SANDOZ laboratories, Paris)

The β-blockers are used for the treatment of cardiac disorders, [22] (hypertension, angina prectoris, cardiac arrhytmias). The medical effects are well proved by physicians [18]. The main difficulty comes from the nonlinearity of the phenomenon.

We relate the blood concentration of such drug and the medical effect (Fig.8). The experimental data were given by Servier laboratories (cardiologic division) and are relative to a drug called "Tertatolol".

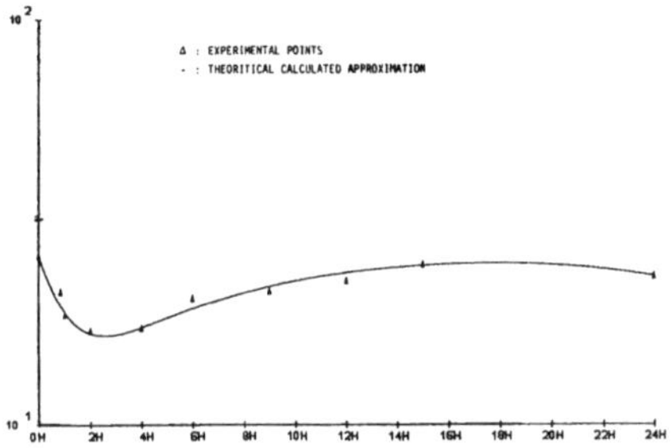

Figure 8. Simulation of the effect DP of the drug. The vertical axis represents the effect of the β-blockers (TERTATOLOL) in μg/ℓ. The horizontal axis is associated to the time in hours.

In (Fig.9) we have a numerical treatment of cardiac diseases using optimal control theory. The experimental data were given by Sandoz (Mrs Guerret and Lavene) [30], [31]. For u(control function) going to 6.5844 the drug concentration x_1 tends to a constant A = 18.

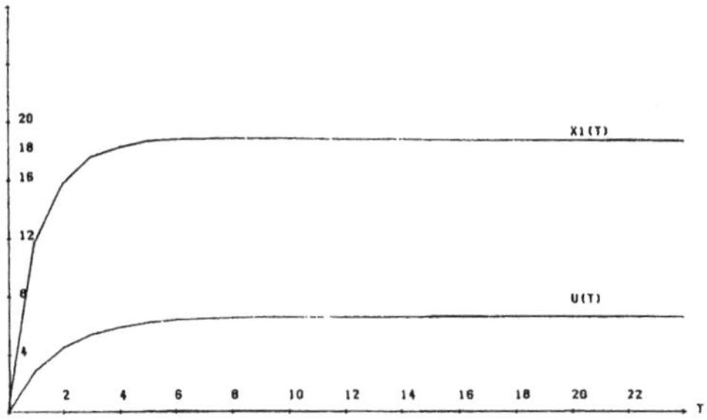

Figure 9. The results of optimal control theory applied to linear compartmental models. The horizontal axis associated to the time in hours.

9. CONCLUSIONS

The present literature is very poor in papers treating of optimization in biology. Some too scarce attempts exist [32] but they consider generally the linear case. Our aim was to treat mainly the nonlinear case and to give some simple and efficient numerical methods for solving optimal control problems. We saw that optimization allows to act on the biological system by minimizing some optimal criterion. Therefore optimal policies may be defined as for instance when using some drugs. In that case doses and times of absorption may be mathematically calculated. In some cases an analytic solution is available while in other cases only a numerical solution can be proposed.

In our study the numerical optimization methods play a great role and specially an original technique based on a "space filling curve" idea (Alienor technique) : Modelling is not sufficient by itself. The most important property of modelling is the possibility of acting on the biological system by mean of a control function. This action can be done so as to minimize some biological criterion.

REFERENCES

[1] S.L. BEAL. 'Computation of the Explicit Solution to the Michaelis-Menten Equation'. *J.Pharmacokin. Biopharm.* **11** : 641-657, (1983).

[2] G. CHEN, J.F. GROSS. 'Pharmacokinetics of Drugs Subject to Enterohepatic Circulation'. *J.Pharmaceutical Sciences.* **68**, n° 6 : 792-794, (1979).

[3] Y. CHERRUAULT. *Biomathématiques.* Coll. Que sais-je ?. P.U.F. Paris, (1983).

[4] Y. CHERRUAULT, M. GUERRET. 'Parameters identification and optimal control in Pharmacokinetics'. *Acta Applicandae Mathematicae* **1**, 105-120, (1983).

[5] Y. CHERRUAULT, A. GUILLEZ. 'A simple method for optimization'. *Kybernetes.* **12**, n° 1 : 59-63, (1983).

[6] Y. CHERRUAULT, A. GUILLEZ. 'L'obtention de tous les minima et du minimum absolu des fonctions de plusieurs variables : Aliénor'. *Cybernetica.* **XXVII**, n° 1, (1985).

[7] Y. CHERRUAULT, I. KARPOUZAS, A. GUILLEZ. 'Une nouvelle méthode numérique pour la résolution d'équations aux dérivées partielles issues de la biologie'. *Biosciences.* **4**, n° 4 : 89-92, (1985).

[8] Y. CHERRUAULT, I. KARPOUZAS, J.F. PROST, A. GUILLEZ. 'Numerical treatment of cardiac diseases using optimal control theory'. (To appear in *Advances in Mathematics and computers in Medicine).*

[9] Y. CHERRUAULT, J.F. PROST, I. KARPOUZAS, M. HOTCHEN. 'Mathematical modelling according for discrepancies between blood levels and pharmacodynamic effects of a drug. Application to the therapeutic optimization of Tertatolol'. (To submit in *Inter. J. of bio-medical computing).*

[10] Y. CHERRUAULT, V.B. SARIN. 'General treatment of linear mammillary models'. *Inter. J. of bio-medical computing.* **16**, n° 2 : 119-126, (1985).

[11] A. FIACCO, G. McCORNICK. *Nonlinear programming : Sequential unconstrained minimization techniques.* J. Willey, (1968).
[12] M. FORTIN, R. GLOWINSKI. *Méthodes de Lagrangien augmenté. Applications à la résolution numérique de problèmes aux limites.* Méthodes mathématiques de l'informatique, 9, Dunod, Paris, (1982).
[13] M. GIBALDI, D. PERRIER. *Pharmacokinetics.* M. Dekker, New York, (1975).
[14] K. GODFREY. *Compartmental models and their application.* Acad. Press. (1983).
[15] K. GODFREY, W. FITCH.'The Deterministic Identifiability of Nonlinear Pharmacokinetic Models.' *J. Pharm. and Biopharm.* 12, n° 2 : 177-191, (1984).
[16] K. GODFREY, W. FITCH.'On the identification of Michaelis-Menten. Elimination Parameters from a single Dose-Response Curve. *J. Pharm. and Biopharmaceutics.*' 12, n° 2 : 193-221, (1984).
[17] D. GREENSPAN. *Numerical Solutions of Nonlinear Differential Equations.* J. Willey, Madison, (1967).
[18] G. JOHNSON, G. JG. REGARTH.'Clinical Pharmacokinetics of B-Adrenoreceptor Blocking Drugs Clinical Pharmacokinetics.' 1 : 233-263, (1976).
[19] D.S. JONES, B.D. SLEEMAN. *Differential equations and mathematicaly biology.* G. Allen. Unwin, (1983).
[20] I. KARPOUZAS.'Une méthode d'optimisation globale appliquée à un problème de mécanique ventilatoire'. *Kybernetes.* 14, n° 4 : 253-260, (1985).
[21] I. KARPOUZAS, Y. CHERRUAULT, J.F. PROST.'Modélisation de l'action des médicaments. Définition de thérapeutiques optimales'. (To appear in *Berichte der Arbeitsgruppe Mathematisierung der GhK).*
[22] I. KARPOUZAS, Y. CHERRUAULT, J.F. PROST, A. GUILLEZ.'Control and optimization of drug action'. (To appear in *Proceedings of Orwellian Synposium and international conference on systems research informatics and cybernetics.* Baden-Baden. 1-5 August, (1984).
[23] A.W. KELMAN, B. WHITING, S.M. BRYSON.'OPT. A package to computer programs for parameter optimization in clinical pharmacokinetics'. *Br. J. Clin. Pharmac.* 14, : 247-256, (1982).
[24] J.L. LIONS. *Contrôle optimal de systèmes gouvernés par des équations aux dérivées partielles.* Dunod, Paris, (1968).
[25] T.M. LUDDEN, J.P. ALLAN, W.A. VALUTSKY. et al.'Individualization of phenytoin dosage regimens'. *Clin. Pharmacol. Ther.* 21 : 287-293, (1977).
[26] P.V. PEDERSEN, R. MILLER.'Pharmacokinetics and Bioavailability of cimetidine in Humans'. *J. of Pharm. Sciences.* 69, n° 4 : 394-398, (1980).
[27] Y. PLUSQUELLEC, J.L. STEIMER.'An Analytical Solution for a class of Delay. Differential Equations in Pharmacokinetic Compartment Modelling'. *Math. Biosciences.* 70, : 39-56, (1984).
[28] L. SCHWARTZ. *Théorie des distributions.* Hermann, Paris, (1966).
[29] L.B. SHEINER.'Analysis of pharmacokinetic Data using Parametric Models - 1 : Regression Models'. Tutorial section. *J.of Pharma. and Biopharm.* 12, n° 1, (1984).

[30] B. SOME. *Identification, contrôle optimal et optimisation dans les systèmes d'équations différentielles compartimentales*. Thèse de 3ème cycle. Université Paris 6, (1984).
[31] B. SOME, Y. CHERRUAULT. 'Optimisation et contrôle optimal en pharmacocinétique'. (To appear in *Biomedicine and Pharmacotherapy)*.
[32] G.W. SWAN. *Application of optimal control theory in biomedicine*. M. Dekker, (1984).
[33] D.D.M. TONG, C.M. METZLER. 'Mathematical properties of compartments models with Michaelis-Menten type elimination'. *Math. Biosc.* 48 : 293-306, (1980).
[34] J. WAGNER. *Fundamentale of clinical pharmacokinetics*. Drug Intelligence, (1975).

Heat and mass transfer in quasi-steady ground freezing processes (*)

A. Fasano - M. Primicerio

Abstract. A model for quasi-steady ground freezing processes is studied with the aim of obtaining simple quantitative information about the water discharge, the location of the freezing front and of possible ice lenses. The model is derived from the one proposed in ref. [2] and based on ice-water thermodynamical equilibrium, introducing some approximations concerning e.g. the averaged thermal conductivity and the freezing characteristic curve.
Some of these approximations are rather crude, but the formulas obtained are very simple, physically expressive and still within the general degree of accuracy of the starting dynamical model.

(*) Work partially supported by USA-ERO Contract N.DAJA45-83-C-0053

1. Introduction

Freezing of saturated porous media is a complex phenomenon for which many mathematical models have been proposed. In a recent review payer [1] they were grouped in two main classes. In the first class, a sharp front separates a totally frozen region from the unfrozen part, while the second class is characterized by coexistence of water and ice in the so-called frozen-fringe.

In the latter case, applying the laws of thermodynamical equilibrium between water and ice seems to provide a satisfactory explanation of the water transport mechanism accompanying freezin processes. In [2] this goal was accomplished assuming a minimal number of empirical relationships between the quantities involved

In this paper we will give such relationships a specific mathemat form which reduces the problem to a free boundary problem and provides easy quantitative information, at least for the case of quasi-steady freezing processes.

The basic concepts of the model are the following. We will consider a planar geometry, i.e. one-dimensional freezing of a horizontal saturated porous layer, z denoting the vertical coordinate, positive upwards ($z=0$ at the bottom of the layer).

(a) Coexistence of water and ice in a frozen fringe.

Denoting by v the volumetric water velocity and by w the fraction of pore volume occupied by water, Darcy's law takes the form (one-dimensional vertical flow)

$$(1.1) \qquad v = -\frac{\tilde{D}(w)}{\rho g}\left(\frac{\partial p}{\partial z} + \rho g\right), \qquad g = 9.8\ \text{m/sec}^2,$$

where ρ and p denote water density and pressure, while $\tilde{D}(w)$ represents the hydraulic conductivity of the medium constituted by ground grains and (if $w<1$) ice particles. According to [2], $D(w)$ is approximated by

$$D(w) = D_o w^7 \quad , \tag{1.2}$$

typical values of D_o for fine sands ranging from 10^{-5} to 10^{-3} m/sec (see [3]).

(b) Thermodinamical equilibrium in the frozen fringe.

In the region where $0 < w < 1$ Clapeyron law applies, i.e.

$$\frac{p}{\rho} - \frac{p_i}{\rho_i} = \frac{\lambda\theta}{T_o} \quad , \qquad \theta = T - T_o \quad , \tag{1.3}$$

where p_i, ρ_i denote pressure and density of the ice, λ is the latent heat, $T(°K)$ is the equilibrium temperature and $T_o(°K)$ is the melting point.
In (1.3) and in the sequel all pressures are referred to atmospheric pressure (gauge pressures).

(c) Ice-water interfacial effects.

A pressure jump between the two phases, due to an interfacial effect, is accounted for by writing the equation

$$p_i - p = \omega\tilde{\psi}(w) \quad , \tag{1.4}$$

where ω is the ice-water surface tension and $\tilde{\psi}$ is an empirically determined function. In the spirit of the Kelvin-Laplace law, $\tilde{\psi}$ is some averaged curvature of the ice- water interface.

A typical profile of $\tilde{\psi}$, taken from [2], is shown in fig. 1.

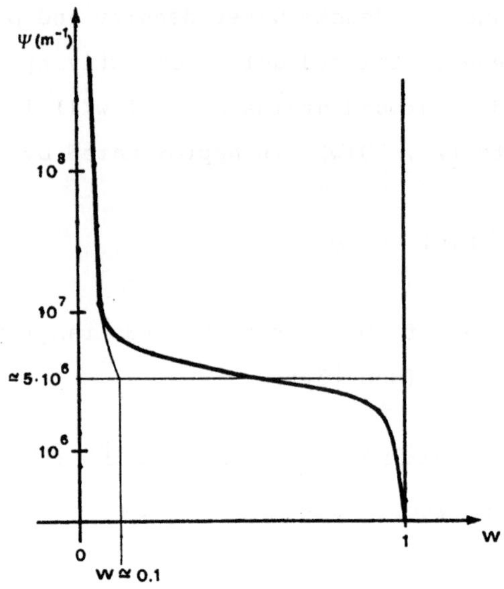

Fig.1 Freezing characteristic curve (thick line experimental).

The analysis performed in this paper stems from the idea of approximating $\tilde{\psi}(w)$ in such a way that

(i) the main information carried by $\tilde{\psi}$ is preserved (namely the presence of a flat branch around some value ψ_o),

(ii) the approximation is reasonable, considering the uncertainty both in determination of $\tilde{\psi}$ and in the empirical law (1.2

(iii) a series of calculations can be performed in the (quasi) steady case, leading to a set of very simple formulas for all the relevant unknown quantities of the problem.

The proposed approximation is

$$(1.5) \qquad \psi(w) = \begin{cases} \psi_\circ \quad , & w_\circ \leq w < 1 \; , \\ \psi_\circ \; (w_\circ/w)^2 \quad , & 0 < w \leq w_\circ \quad . \end{cases}$$

We consider an example with the following specific values: $w_\circ = 0.1$, $\psi_\circ = 5 \cdot 10^6 \; m^{-1}$. However, the choice of a Padé approximation of ψ for $0 < w < w_\circ$ is not crucial and is made here to make calculations simpler.

The coupled heat and mass transfer equations are obtained inserting (1.1)-(1.4) in the equations expressing mass and energy conservation. Heat transfer due to water flow will be neglected (this is largely justified when forced convection is absent see [6]) and ice in the frozen fringe below the lowest ice lens will be supposed to be at rest. The latter assumption is consistent with the fact that (i) in the spirit of (1.1)-(1.2) ice is seen by flowing water as a component of the porous matrix, and (ii) Darcy's low (1.1) is written in a frame of reference in which solid particles are at rest.

An alternative form of the system (1.3),(1.4), frequently used in the sequel is

$$(1.6) \qquad p = - \left[\omega\psi(w) + C\theta \right] / \delta \; ,$$

$$(1.7) \qquad p_i = - \left[(1-\delta) \, \omega\psi \, (w) + C\theta \right] / \delta \; ,$$

where

$$(1.8) \qquad \delta = \frac{\rho - \rho_i}{\rho} \quad , \quad C = \frac{\rho_i \lambda}{T_\circ}$$

and $\delta \neq 0$.(*)

When $\delta = 0$, then (1.3),(1.4) provide a relationship between θ an w

$$(1.9) \qquad \theta = -\frac{T_o \omega}{\rho\lambda} \psi(w) = -\frac{\omega}{C} \psi(w) < 0 .$$

Although a steady state situation is possible strictly speaking only if no water flow occurs through the medium (since such flow is always related to the evolution of an ice lens, possibly at the top of the layer), nonetheless the perturbation induced by slowly growing lenses is generally so small that it makes sense to look for (almost) steady states. Moreover, if the thermal gradient changes sufficiently slowly in time, it can be interesting to describe the evolution of the system through a sequence of steady states.

The discontinuity introduced for ψ makes a freezing front appear (instead of a narrow region of sharp variation for z). Referring to steady states, we will find formulas for the determination of the location of the freezing front. Moreover, we will express in terms of the data both the water discharge through the medium (and consequently the growth rate of the lowest ice lens), and the location of the ice lenses. In addition we will get very simple expressions for such quantities when $\delta = 0$ (i.e. when the density jump accross the phases is neglected) and we will describe the influence of δ as a

---

(*) Although we are referring to the water-ice system, we want to derive formulas valid in the general case and emphasizing the dependence on δ. This will turn out to be useful even when dealing with the water-ice system itself, since it gives an idea of the error made in using simpler formulas with $\delta = 0$.

perturbation (having the form of easily computable power series expansions, valid within some range of δ, including negative values).

This procedure provides a very clear and at least qualitatively reliable interpretation of the way the various parameters affect the water flow. At the same time it answers same non-intuitively solvable questions.

Quite recently a quasi-static approach for ground freezing has been considered in [7], also starting from the evolutive model of [2] but proceeding through different approximations. We stress that our approach is aimed at obtaining explicit approximating formulas.

2. The basic equations.

We consider a saturated porous layer between two horizontal planes $z = 0$ (the bottom), $z = L$ (the top).

As a preliminary we will assume that no ice lens is present inside the layer and that all water possibly emerging from the medium freezes instantly. Thus we impose the following time independent boundary conditions (the perturbation introduced by the ice layer forming at the top is absolutely negligible)

$$(2.1) \qquad \theta = \theta_a < 0 \,, \quad p_i = p_a \,, \quad \text{at} \quad z = L \,,$$

$$(2.2) \qquad \theta = \theta_b > 0 \,, \quad p = p_b = p_a + \rho g L \,, \quad \text{at} \quad z = 0$$

(in laboratory tests the water pressure at the bottom can be controlled from outside).

A remarkable simplification is obtained if one assumes that the thermal conductivity is constant throughout the medium. Although such a quantity depends at least on w, we note that this dependence is largely mitigated by the fact the ice-water system occupies only a fraction of the total volume. Such an approximation seems quite reasonable, since the model contains more significant sources of uncertainty. Moreover we remark that removing this simplification implies nothing but formal complications and does not alter substantially any of the conclusions of the present analysis.

Therefore we will write

$$(2.3) \qquad \theta(z) = \theta_b - G z \,, \qquad 0 \le z \le L \,,$$

where

$$G = \frac{\theta_b - \theta_a}{L} \quad . \tag{2.4}$$

Below the freezing front $z = s$ (still to be determined) pressure is a linear function

$$p(z) = p_b - (q + \rho g)z \quad , \quad 0 < x < s \ , \tag{2.5}$$

where $q \geq 0$ is an unknown constant related to water volumetric velocity by the formula

$$v = \frac{D_o}{\rho g} q \quad . \tag{2.6}$$

Imposing again $\frac{dv}{dz} = 0$ above the freezing front, by the continuity of v accross $z = s$ we get

$$- w^7 \left(\frac{dp}{dz} + \rho g \right) = q \quad , \quad s < x < L \ , \tag{2.7}$$

where (1.2) and (2.6) have been used.

3. The location of the freezing front.

Matching the values of pressure for $z \to s-$ (deduced from (2.5)) and for $z \to s+$ (use (1.6),(2.3) and the fact that $\psi \to \psi_o$), we get for $\delta \neq 0$

$$p_b - (q + \rho g)s = -\omega\psi_o/\delta - C\left\{\theta_b - \frac{\theta_b - \theta_a}{L} s\right\}/\delta \,, \tag{3.1}$$

i.e.

$$\frac{s}{L} = \frac{\theta_b - \theta_o + p_b\delta/C}{\theta_b - \theta_a + L(\rho g + q)\delta/C} \,, \tag{3.2}$$

where

$$\theta_o = -\frac{\omega\psi_o T_o}{\lambda \rho_i} = -\frac{\omega\psi_o}{C} \,. \tag{3.3}$$

Of course (3.2) makes sense only if the r.h.s. is less than 1. This is guaranteed by the condition $\theta_o - \theta_a > (p_a - Lq)\delta/C$.

In practical cases, it will be sufficient to check that

$$\theta_o > \theta_a \,. \tag{3.4}$$

It can be seen that (3.2) is an $O(\delta/C)$ perturbation of the value

$$\frac{s_o}{L} = \frac{\theta_b - \theta_o}{\theta_b - \theta_a} \,, \tag{3.5}$$

more precisely,

(3.6) $$\frac{s}{L} = \left(\frac{s_\circ}{L} + \frac{p_b \delta/C}{\theta_b - \theta_a}\right) \Big/ \left(1 + \frac{L(q+\rho g)\delta/C}{\theta_b - \theta_a}\right) ,$$

or to the first order in δ/C (= $7.35\cdot 10^{-8}$ for water, MKS units)

(3.7) $$\frac{s}{L} = \frac{s_\circ}{L} + \frac{\delta/C}{\theta_b - \theta_a} \left[p_b - s_\circ(\rho g + q) \right] ,$$

where again q is expected to be negligible (as well as ρg).

The reduced influence of q on the determination of s is of some importance, since q is mostly affected by the approximations of $\tilde{D}$ and of $\tilde{\psi}$ introduced in the model.

At this point we note the following.

(i) If $\delta = 0$, then (1.5),(1.9), and (3.3) give

(3.8) $$\theta = \theta_\circ\, w_\circ^2/w^2 \quad , \quad \text{if} \quad \theta < \theta_\circ \; , \quad w = 1 \quad \text{otherwise,}$$

showing that $\theta_\circ$ represents the temperature at the freezing front, which is located at $z = s_\circ$.

(ii) The condition (3.4) has now a clear physical meaning.

(iii) (3.7) shows that the correction introduced by δ on the dimensionless quantity $\frac{s}{L}$ can be very small in practical cases (less than 10^{-2} for water, $p_b \cong$ atmospheric pressure, $\theta_b - \theta_a \cong 1$ °C).

We finally remark that in [4] an empirical relationship $w=w(\theta)$ of the same type as (3.8) is assumed, in place of the thermodynamical equations leading to the introduction of the function $\tilde{\psi}(w)$.

However in [4] the densities of the two phases are kept distinct.

4. The water discharge (existence of the solution).

In order to determine the constant q first we have to integrate (2.7). Recalling (1.6) and (2.3), we write

$$w^7\left\{\omega \frac{d\psi}{dw}\frac{dw}{dz} - CG - \rho g\delta\right\} = q\delta \quad , \quad s < z < L . \tag{4.1}$$

From now on we neglect the gravity term everywhere, since it is actually very small in comparison to other terms and it carries nothing but some more complicate algebra.

We set

$$\eta = q\, w_a^{-7}\, \frac{\delta}{CG} \quad , \tag{4.2}$$

where $w_a = w(L)$ is obtained from (1.7),(2.1)

$$w_a = w_\circ \left[\frac{\theta_\circ(1-\delta)}{\theta_a + p_a\delta/C}\right]^{\frac{1}{2}} \tag{4.3}$$

(in practical cases note that (4.3) makes sense because of the assumption (3.4)).

Integrating (4.1) between $z = s$ and $z = L$, we get

$$2\,\frac{-\theta_\circ}{\theta_b - \theta_a}\,\frac{w_\circ^2}{w_a^2}\int_1^{w_\circ/w_a} \frac{y^4}{y^7+\eta}\,dy = 1 - \frac{s}{L} \tag{4.4}$$

(in view of the fact that we want to allow δ to assume also negative values, the unknown quantity η in (4.4) has to be looked for in the interval $(-1,+\infty)$).

Of course equation (4.4) can be solved numerically (the integral can also be calculated explicitly), but we will use a different procedure (see Sec.5), leading to a simple expression of the solution, exhibiting its dependence on δ. First of all let us show that a solution exists.

We use (3.2), neglecting ρg, to write (4.4) in the form

$$\alpha(\eta,\delta) = \beta(\eta,\delta) , \tag{4.5}$$

where

$$\alpha(\eta,\delta) = 2\int_1^{w_\circ/w_a} \frac{y^4}{y^7+\eta}\, dy \tag{4.6}$$

and

$$\beta(\eta,\delta) = (1-\delta)\frac{\theta_b-\theta_a}{-\theta_a-p_a\delta/C}\,\frac{1}{1+\eta w_a^7}\left\{\frac{\theta_\circ-\theta_a}{\theta_b-\theta_a} - \frac{p_a\delta/C}{\theta_b-\theta_a} + \eta w_a^7\right\} . \tag{4.7}$$

We prove that (4.5) has a unique solution $\eta \in (-1,+\infty)$. First we consider the case $\delta \neq 0$. The case $\delta = 0$ will be considered separately with the aim of getting the general solution $(\delta\neq 0)$ as a perturbation of the $\delta = 0$ solution, which has a particularly simple expression.

We calculate

$$\alpha_\circ(\delta) \equiv \alpha(0,\delta) = 1-\left(\frac{w_a}{w_\circ}\right)^2 = 1-(1-\delta)\frac{\theta_\circ}{\theta_a+p_a\delta/C} , \tag{4.8}$$

$$\beta_\circ(\delta) \equiv \beta(0,\delta) = (1-\delta)\left\{1-\frac{\theta_\circ}{\theta_a+p_a\delta/C}\right\} . \tag{4.9}$$

Note that

$$\alpha_\circ(\delta) - \beta_\circ(\delta) = \delta \tag{4.10}$$

and that

$$(4.11) \qquad \alpha'(\eta) < 0 \ , \quad \lim_{\eta\to+\infty} \alpha(\eta) = 0 \ , \quad \lim_{\eta\to-1+} \alpha(\eta) = +\infty \ ,$$

$$(4.12) \qquad \frac{\partial\beta}{\partial\eta} > 0 \qquad \text{(it only requires } \theta_b - \theta_\circ + p_a\delta/C > 0\text{)}.$$

Therefore it follows that there exists a unique solution $\eta(\delta)$ of (4.5) having the same sign as δ, so that the corresponding value of q is always positive.

In the case $\delta = 0$, using (3.8) (together with (1.4),(2.1), (2.5)) equation (2.7) can be integrated as follows

$$(4.13) \qquad -\omega\psi_\circ(w_\circ/w_a)^2 + qs_\circ = -qw_\circ^{-7}\int_{s_\circ}^{L}\left[\theta(z)/\theta_\circ\right]^{7/2}dz \ .$$

Recalling (2.3) it follows that

$$q\left\{s_\circ - \frac{2}{9}\frac{\theta_\circ}{G}w_\circ^{-7}\left[\left(w_\circ/w_a\right)^9 - 1\right]\right\} = \omega\psi_\circ\left(w_\circ/w_a\right)^2$$

and, neglecting $s_\circ$,

$$(4.14) \qquad q = \frac{9}{2}\frac{G\omega\psi_\circ}{|\theta_\circ|}\frac{w_a^7}{1-(\theta_\circ/\theta_a)^{9/2}} \ .$$

When $(\theta_\circ/\theta_a)^{9/2} << 1$, we obtain the value

$$(4.15) \qquad q \cong q_\circ = \frac{9}{2}\, G C w_\circ^7\, (\theta_\circ/\theta_a)^{7/2}$$

(see also (3.3),(3.8), which agrees substantially with the formula proposed in [5]).

In our example $\theta_\circ \cong -0.14°C$, $C \cong 1.1\cdot10^6$ MKS units, thus $q_\circ \cong 0.5\, G\, (\theta_\circ/\theta_a)^{7/2}$ (with G in °C/m the result is in MKS units), which is several magnitude orders smaller that the already neglected gravimetric pressure gradient ρg.

5. The water discharge (determination of the solution).

We want to find $q(\delta)$ from (4.5) as a power series having (4.15) as the zero order term. This means that the approximation underlying this procedure is

$$(5.1) \qquad (\theta_\circ/\theta_a)^{9/2} \ll 1 \; .$$

First of all we want to remark that the approximation

$$(5.2) \qquad \beta(\eta,\delta) \cong \beta_\circ(\delta)$$

is valid in practical cases.

Let us prove (5.2). If $\delta < 0$ then $-1 < \eta < 0$ and $|\eta w_a^7| < w_\circ^7$.

On the other hand, if $\delta > 0$, setting $a = \delta/\beta_\circ$ we find that

$$\alpha\left[a\left(\frac{w_\circ}{w_a}\right)^7, \delta\right] < 2\int_1^{w_\circ/w_a} \frac{y^4}{(1+a)y^7}\, dy \; .$$

Therefore, by virtue of (4.10),

$$\alpha\left[a\left(\frac{w_\circ}{w_a}\right)^7, \delta\right] < \frac{\alpha_\circ(\delta)}{1+a} = \beta_\circ(\delta) \; .$$

Hence

$$\eta w_a^7 < \frac{\delta}{\beta_\circ} w_\circ^7 \; .$$

Going back to (4.7), we can specify what is the range of θ_a in which (5.2) is valid, and the limitation imposed by (5.1) guarantees that ηw_a^7 is negligible in (4.7). Hence (5.2) is proved.

Next we remark that δ enters the function $\alpha(\eta,\delta)$ through the ratio w_o/w_a. Since

$$D^n_\eta \, \alpha(0,\delta) = (-1)^n \frac{2n!}{7n+2} \left[1 - \left(\frac{w_a}{w_o}\right)^{7n+2} \right] ,$$

we see that if $(w_a/w_o)^9 \ll 1$ the dependence on δ is lost in all the derivatives of α, and we can write

$$(5.3) \qquad \alpha(\eta,\delta) = \alpha_o(\delta) + 2 \sum_{n=1}^{\infty} (-1)^n \frac{\eta^n}{7n+2}$$

Thus, recalling (4.10), equation (4.5) becomes

$$(5.4) \qquad 2 \sum_{n=1}^{\infty} (-1)^{n-1} \frac{\eta^n}{7n+2} = \delta ,$$

the series on the l.h.s. being convergent for $\eta \in (-1,1]$.

Inserting in (5.4) the tentative expansion

$$(5.5) \qquad \eta(\delta) = \sum_{K=1}^{\infty} \lambda_K \, \delta^K ,$$

one finds recursive formulas for the coefficients λ_K :

$$\lambda_1 = \frac{9}{2} , \quad \lambda_2 = \frac{9}{2} \cdot \frac{1}{8} \lambda_1^2 ,$$

$$\lambda_3 = \frac{9}{2} \left\{ \frac{1}{4} \lambda_1 \lambda_2 - \frac{2}{23} \lambda_1^3 \right\} ,$$

$$\lambda_4 = \frac{9}{2} \left\{ \frac{1}{8}(\lambda_2^2 + 2\lambda_1\lambda_3) - \frac{6}{23} \lambda_1^2 \lambda_2 + \frac{1}{15} \lambda_1^4 \right\} ,$$

$$\lambda_5 = \frac{9}{2} \left\{ \frac{1}{4}(\lambda_1\lambda_4 + \lambda_2\lambda_3) - \frac{6}{23} (\lambda_1\lambda_2^2 + \lambda_1^2\lambda_3) + \frac{4}{15} \lambda_1^3 \lambda_2 - \frac{2}{37} \lambda_1^5 \right\} , \; \ldots\ldots$$

Now we can say how q_o is perturbed by δ. Setting

$$\frac{q-q_o}{q_o} = \kappa(\delta) \quad , \tag{5.6}$$

from (4.2),(4.5) we have

$$\kappa(\delta) = \frac{2}{9}\frac{\eta}{\delta} - 1 \, , \tag{5.7}$$

i.e. (at least formally)

$$\kappa(\delta) = \sum_{K=1}^{\infty} \frac{2}{9} \lambda_{K+1} \delta^K = \tag{5.8}$$

$$= 2.51\delta + 4.99\,\delta^2 + \ldots$$

It has to be noticed that δ is an analytic function of η for $|\eta| < 1$ in the complex plane and that the series (5.4) is nothing but

$$\delta = 2\int_1^{\infty} \left(\frac{1}{y^3} - \frac{y^4}{y^7+\eta}\right) dy = 1 - 2\int_1^{\infty} \frac{y^4}{y^7+\eta}\, dy \ . \tag{5.9}$$

Thus when η goes from -1 to 1 on the real axis $\delta'(\eta)$ never vanishes (and δ increases on the real axis from $-\infty$ to some $\delta_o \cong 0.15$)

Moreover, setting $\eta = \xi + i\zeta$.

$$\operatorname{Im} \delta'(\eta) = 2 \operatorname{Im} \int_1^{\infty} \frac{y^4}{(y^7+\eta)^2}\, dy = \tag{5.10}$$

$$= -4\zeta \int_1^{\infty} \frac{(y^7+\xi)\,y^4}{\left|(y^7+\xi)^2+\zeta^2\right|^2}\, dy$$

never vanishes for $\zeta \neq 0$ in the circle of radius 1.

Therefore $\delta'(\eta) \neq 0$ in the circle of radius 1 and the inverse function $\eta(\delta)$ is also analytic in the image of $|\eta|<1$.

As an conclusion we can say that for values of $|\delta|$ small enough the series (5.8) is convergent. Back to real values of δ we remark the following.

(i) the relative correction to q_o is $\cong 2.5\,\delta$ and therefore it has the same sign as δ,

(ii) higher order corrections are absolutely negligible,

(iii) the coefficients in (5.4) and (5.8) depend only on the shape chosen for the functions ψ,D.

6. The ice pressure

At this point $p_i(z)$ is given by (1.7), once w(z) has been found from the equation

(6.1) $$2\int_{w/w_a}^{w_\circ/w_a} \frac{y^4}{y^7+\eta}\,dy = \left(\frac{w_a}{w_\circ}\right)^2 \frac{\theta_b-\theta_a}{-\theta_\circ}\left(\frac{z}{L}-\frac{s}{L}\right), \quad z\in(s,L),$$

which is the obvious modification of (4.4). Equation (6.1) can be integrated in the form

(6.2) $$2\int_{w/w_a}^{w_\circ/w_a} \sum_{n=0}^{\infty}(-1)^n\eta^n\, y^{-7n-3}\,dy = \left(\frac{w_a}{w_\circ}\right)^2 \frac{\theta_b-\theta_a}{-\theta_\circ}\left(\frac{z}{L}-\frac{s}{L}\right),$$

with $|\eta| < 1$. At the zeroth order in η (i.e. for $\delta = 0$) it gives

(6.3) $$\left(\frac{w_\circ}{w}\right)^2 = \frac{z}{s_\circ}\left(1-\frac{\theta_b}{\theta_\circ}\right)+\frac{\theta_b}{\theta_\circ}, \qquad z\in(s_\circ,L),$$

which is identical to (3.8).

The study of the profile of $p_i(z)$ is of great importance for the possible formation of an ice lens as illustrated in [2]. Some information can be easily obtained if we differentiate (1.7), using (2.7) with no gravity term

(6.4) $$\frac{dp_i}{dz} = -(1-\delta)\,q\,w^{-7} + CG,$$

(6.5) $$\frac{d^2p_i}{dz^2} = 7(1-\delta)\,q\,w^{-8}\frac{dw}{dz} < 0,$$

since $\frac{dw}{dz} < 0$ as it is easily seen from (6.1). Therefore $\frac{dp_i}{dz}$ decreases from the value

$$- (1-\delta) \quad q \, w_o^{-7} \quad + \quad CG$$

(assumed at z=s) to the value

$$- (1-\delta) \quad q \, w_a^{-7} \quad + \quad CG$$

(assumed at z=L).

As a consequence p_i has a maximum between z=s and z=L if

(6.6a) $$\frac{q \, w_o^{-7}}{C G} < \frac{1}{1-\delta} \qquad \text{and}$$

(6.6b) $$\frac{q w_a^{-7}}{C G} > \frac{1}{1-\delta} \quad .$$

Owing to (4.2) the second inequality is equivalent to $\frac{\eta}{\delta} > \frac{1}{1-\delta}$.

If $\delta > 0$, a lower estimate for η is

(6.7) $$\eta > \frac{9}{2} \delta \qquad (\delta > 0) ,$$

which is easily derived from (5.4) (such a series has alternating signs when $\eta > 0$). Hence (6.6b) is satisfied if $0<\delta<\frac{7}{9}$. Using the first few terms of (5.5), it can be seen that the same is true e.g. for $\delta > -0.1$.

The inequality (6.6a) can be written as

$$|\eta| \; (w_a/w_o)^7 < \frac{|\delta|}{1-\delta}$$

and using again (5.5) it is implied by

$$(6.8) \qquad w_a < w_\circ \left\{\frac{2}{9}\,\frac{1}{1-\delta}\right\}^{1/7}, \ \delta \leq 0 \ ; \quad w_a < w_\circ \left\{\frac{7}{72}\,\frac{1}{1-\delta}\right\}^{1/7}, \ 0 < \delta < \delta_\circ$$

which shows very little dependence on δ (practically $w_a < 0.81 w_\circ$) and can be read as a condition on θ_a, recalling (4.3).
The maximum $\hat{p}_i$ of p_i is assumed for the value of w satisfying

$$q\,\hat{w}^{-7} = \frac{CG}{1-\delta}$$

(see (6.4)), i.e.

$$(6.9) \qquad \hat{w} = w_a \left[\frac{9}{2}\,(1+\kappa(\delta))\,(1-\delta)\right]^{1/7} \cong w_a \left[\frac{9}{2}\,(1+1.5\delta)\right]^{1/7},$$

again showing little dependence on δ. Indeed $\hat{w} \cong 1.24\ w_a$, the correction factor being $\cong 1 + 0{,}21\delta$.

The corresponding location of the maximum, $z = \hat{z}$ can now be found from (6.1) or (6.2). Temperature $\hat{\theta} = \theta(\hat{z})$ is given by (2.3). For $\delta = 0$ (3.8) together with (6.9) can be used to determine $\hat{\theta}_\circ$

$$(6.10) \qquad \hat{\theta}_\circ = \theta_a (w_a/\hat{w})^2 = \left(\frac{2}{9}\right)^{2/7} \theta_a \cong 0.65\ \theta_a ,$$

so that from (2.3) the location $\hat{z}_\circ$, of the maximum can be found

$$(6.11) \qquad \frac{\hat{z}_\circ}{L} = \frac{\theta_b - (2/9)^{2/7}\theta_a}{\theta_b - \theta_a} .$$

Also in this case the coefficient $(2/9)^{2/7}$ is related to the shape of $\psi(w)$ and $D(w)$ only.

7. The onset of an ice lens

As illustrated in [1] , the ice pressure is mainly responsible for the formation of ice lenses. It is generally assumed that no ice lens is formed as long as the so called "neutral stress" is less than some value p*, depending on the overburden and on the cohesion of the medium. The neutral stress is defined as a convex combination $(1-\phi(w))p_i + \phi(w)p$, where $\phi(w)$ is the stress partition parameter $(\phi(w) = w^{3/2}$ in [2]).

In what follows p* is assumed to be large enough (more precisely $p^* > p_a + \omega\psi_\circ \cong p_i(s)$) and constant throughout the medium. Moreover we will set $\phi = 0$ in order to simplify our analysis, in view of the small influence of the water pressure $(\phi(w_\circ) \cong 3.4 \cdot 10^{-2})$. Of course this effect could also be taken into account with little additional work.

Thus we have a simple criterium based on the calculus of $\hat{p}_i$ and its comparison with p*.

In this section we will confine our analysis to the case $\delta = 0$. The analysis of the previous section shows how to get the perturbation caused by $\delta \neq 0$.

First of all we integrate (2.7) between s and z in the same way we did to get (4.13). The result is

$$(7.1) \qquad p(z) = p_a - \frac{q_\circ w_\circ^{-7} |\theta_\circ|}{G} \frac{2}{9} \{ (\frac{\theta}{\theta_\circ})^{9/2} - 1 \} .$$

Recalling (1.4), (3.3), (3.8) and (4.15), we obtain $p_i(z)$:

$$(7.2) \qquad p_i(z) = p_a - \omega\psi_\circ (\frac{\theta_\circ}{\theta_a})^{7/2} \{ (\frac{\theta}{\theta_\circ})^{9/2} - 1 \} + \omega\psi_\circ \frac{\theta}{\theta_\circ} .$$

According to (6.8) we assume

$$(7.3) \qquad \theta_a < \left(\frac{9}{2}\right)^{2/7} \theta_\circ \cong 1.54\ \theta_\circ \cong -0.22\,°C, \text{ or } \frac{\theta_\circ}{\theta_a} < 0.65,$$

and we insert (6.10) into (7.2) to get

$$(7.4) \qquad \hat{p}_i = p_a + \omega\psi_\circ\, \Phi(\tau) \quad ,$$

where $\tau = \dfrac{\theta_a}{\theta_\circ}$ and

$$\Phi(\tau) = \frac{7}{9}\left(\frac{2}{9}\right)^{2/7}\tau + \tau^{-7/2} \quad , \quad \tau \geq \left(\frac{9}{2}\right)^{2/7} .$$

It is immediately seen that $\Phi(\tau)$ increases from its minimum $\Phi\left[\left(\frac{9}{2}\right)^{2/7}\right] = 1$ to $+\infty$ in the interval $\left(\frac{9}{2}\right)^{2/7} \leq \tau < +\infty$. Therefore, provided $p^* > p_a + \omega\psi_\circ$, no ice lens appears if $\theta_a > \theta_a^* = \theta_\circ\tau^*$, τ^* being the unique solution of the equation

$$(7.5) \qquad \Phi(\tau) = \frac{p^*-p_a}{\omega\psi_\circ} > 1 \quad .$$

The corresponding location of the basis of the lens is (see (6.11))

$$(7.6) \qquad \frac{z_\circ^*}{L} = \frac{\theta_b - (2/9)^{2/7}\,\theta_a^*}{\theta_b - \theta_a^*} \quad .$$

If e.g. p^* is suffficiently large so that $\Phi(\tau^*) \cong 0.5\ \tau^*$, then $\theta_a^* \cong 2\,\dfrac{p^*-p_a}{\omega\psi_\circ}\,\theta_\circ$ and (7.6) becomes

$$(7.7) \qquad \frac{z_\circ^*}{L} = \frac{\theta_{b/|\theta_\circ|} + 1.30(p^*-p_a)/(\omega\psi_\circ)}{\theta_{b/|\theta_\circ|} + 2(p^*-p_a)/(\omega\psi_\circ)}$$

Remark 7.1 θ_a^* does not depend on L.

Remark 7.2 In the analysis above the width of the lens has been neglected, as well as the fact that the thermal gradient has a jump across the lower face of the lens, according to the Stefan condition. As a matter of fact, the ratio $v\lambda/(kG)$ k being the averaged thermal conductivity, in the case of water for $D_\circ = 10^{-3}$ m/sec, $k \cong 0.5$ W/m °C can be estimated as $5\cdot 10^{-2}\ (\theta_\circ/\theta_a)^{7/2} \ll 1$.

Such simplifications do not have any important influence on the results and the corresponding corrections can be calculated with obvious modifications.

8. Calculating the sequence of ice lenses in a quasi-steady cooling process.

Let us fix θ_b and vary θ_a - e.g. linearly - in order to cool the medium in a sufficiently slow way. Then we apply all the analysis above, assuming that the medium passes through a set of equilibrium states. The error introduced will of course depend on how slow the process is.

When θ_a reaches the value $\theta^{(1)} = \theta_a^*$, the first lens will start forming at the location $z^{(1)} = z_o^*$ given by (7.6), where the temperature has attained some value $\hat{\theta}^{(1)}$.

From this time on all the quantities below the lens can be calculated in the way explained so far, substituting $z=L$ by $z=z^{(1)}$. In particular we will be able to calculate the water flux and the corresponding growth rate of the lens.

As soon as the temperature at $z=z^{(1)}$ drops below θ_a^* (remember Remark 7.1), a new lens is formed at $z=z^{(2)}$ such that $z^{(2)}/z^{(1)} = z^{(1)}/L$ (applying Remark 7.1 again).
This procedure can be continued leading to the formula

$$(8.1) \qquad \frac{z^{(n)}}{L} = \left(\frac{z^{(1)}}{L} \right)^n ,$$

$z^{(n)}$ representing the location of the n-th lens.

Of course (8.1) makes sense only if the cumulative thickness of the ice lenses is much smaller than L.

An interesting feature of (8.1) is the fact that all the dependence on the physical parameters lies in $z^{(1)}$, the power law structure being quite general. A qualitative agreement can be observed with the sequence of lenses shown in [5] on a sample after a freezing test.

References

|1| A.FASANO-M.PRIMICERIO, Freezing of porous media: a review of mathematical models. Proc. German-Italian Symposium "Applications of Mathematics in Technology", V.Boffi, H.Neunzert Eds., Teubner (1978), pp. 288-311.

|2| K.O'NEILL - R.D.MILLER, Numerical solutions for a rigid-ice model of secondary frost heave. CRREL Rep. 82-13 (1982).

|3| J.BEAR, Dynamics of Fluids in Porous Media. American Elsevier (1972).

|4| D.BLANCHARD - M.FREMOND, Soils frost heaving and thaw settlement. ISGF 85, S.Kinosita - M.Fukunda eds., Balkema (1985), vol. I, pp. 209-216.

|5| J.P.G.LOCH, Frost heave mechanism and the role of the thermal regime in heave experiments on Norwegian silty soils. Meddelelse 50 (1977), 5-14.

|6| G.S.TAYLOR - J.N.LUTHIN. Numeric results on coupled heat mass flow during freezing and thawing. Proc. 2nd Conference on soil-water problems in cold regions, Edmonton 1976, pp. 155-172.

|7| J.T.HOLDEN - D.PIPER - R.H.JONES. Some developments of a rigid-ice model of frost heave. Fourth Int. Symposium on ground Freezing, Sapporo 1985, S.Kimonita, M.Fukuda eds., Balkema (1985), vol.I, pp. 93-99.

MAKING UNIVERSITY- AND INDUSTRIAL MATHEMATICS MEET. SOME EXPERIENCES FROM FINLAND

Matti Heiliö
Lappeenranta University of Technology
Box 20
53851 Lappeenranta
Finland

ABSTRACT. In this article an activity called Industrial Mathematics Project is described. The project was started at the Further Education Center of Helsinki University to promote the professional use of mathematics in industry, business life etc. Some experiences gained from the project are reported. Also a list of examples from finnish industries is given where mathematics is being used to solve practical problems.

The pure mathematics era

Up to the present day the prevailing direction in the mathematical culture of Finnish universities has been the so called "pure mathematics", a feature common to most countries, I have learned. The papers of mathematicians have contained prolific amounts of operators, measures, groups, vector spaces, homotopyclasses etc. and impressive theorems about them, but (excluding a few) not a hint of the secular things of real life like industrial processes, electronic components, machines, buildings, forests, birds and flowers.

This phenomenon, sometimes called "the Bourbaki Syndrome" is well known to all of you. At present it seems to be well reported, well understood and under serious undertakings. This symposium is a part of the coming of a new era.

The awakening

Starting in the 70's there has been a mild but distinct feeling of awakening. I see several reasons for this.

1. Frustration about the bad employment rate of mathematicians. Very few people with mathematical education get jobs in industry. There is no such profession as an industrial mathematician. On the other hand there is a flock of competent mathematicians without a permanent academic post.
2. The development of the role of the university - according to current public expectations the university should take a more active

and more practical position in society. This has been reflected in the appearance of so-called further education modules, facilities annexed to universities and aiming to "sell and market" the brainpower of the university to the outside world. These further education centres are dynamic and flexible organizations working at the peripheral area of traditional university activity. They form an ideal environment for new initiatives, mathematics inclusive.

3. The coming of computers. The tremendous computing power has provided a long extension to the range of possibilities in the area of mathematical engineering. The sad thing is that we mathematicians have been too slow to admit it and to take the challenge.

The change first appeared in the corridor talks of mathematics departments. People started to realize the situation and there was a commonly felt need for change. At present I can see in Finnish universities a firm determination to increase and diversify university-industry interplay and also to reshape and develop the teaching of mathematics.

The Mathematics in Industry Project

As a constructive commitment to improve the frustrating situation I made in the year 1982 an initiative to start an activity called the "Industrial Mathematics Project".

I drafted a paper containing an anlysis of the poor state of affairs in the practical use of mathematics and a proposal for a project intended to "improve the utilization of mathematical knowledge in industry, business, public administration etc". I presented the plan to the further education center of Helsinki University. The idea was accepted and I was asked to be the project coordinator.

The idea of the project was to start a multi-step, multi-level communication program between universities and industry and other sections of business life.

The planned levels of activity are

- contact seminars and workshops between university- and industry people
- short further education courses according to the demand encountered
- to act as a mediator for consulting contacts.

Industrial Mathematics Workshop 1984

As a first step in the project we organized a seminar "Industrial Mathematics Workshop" which was held in January 1984.

The purpose of the workshop was

- to demonstrate various applications of mathematical methods in practical situations
- to deliver information about current trends in mathematical research
- to discuss the possibilities of improving university/industry cooperation in joint research.

Participants of the seminar came from mathematical and statistical departments of different universities (25), from industry (36) and from different brances of public administration (18).

The program contained 25 short presentations such as

- case-studies from industry,
- short lectures on topics of relevance with respect to the professional use of mathematics,
- reports describing some current research activities.

The program also contained group discussions, information briefings and software demonstrations. The group discussions were an essential part of the program. Working on a prefabricated list of questions the groups were doing on-site brainwork trying to catch the essential problems, obstacles and possibilities at hand. The topics under study were

1. The need and means for further education in mathematics.
2. How to develop the teaching of mathematics in universities?
3. How university/industry cooperation and service research should be organized?
4. The professional identity of an industrial mathematician.

The "results and suggestions"-chapter of this article reflects the moods and opinions expressed at these discussions and can be considered as a partial (and subjective) summary of the results. The participants of the seminar were very satisfies with the idea and there was general agreement about continuing the program. Right now I am preparing the second "Industrial Mathematics Workshop" which will be held in January 1986. We are also planning to organize some further education courses for industry people in near future.

The course of action

To stress some essential details I want to describe the course of action in some detail.

1. The first thing to do is to find potentially interested people. By picking them one by one we now have a mailing list of approximately 200 names.
2. To the people in our list we mailed a leaflet describing the project plus an inquiry about
   - their interest to participate
   - description of mathematical applications in their company
   - suggestions for the program
   - willingnes to produce a presentation
   - further references to interested people
   - willingnes to allow a visitor to come for a personal interview.
3. We got about 80 answers and about 30 of the companies were ready to allow a visitor. So during last summer I visited about 15 companies staying 2-6 hours in each to discuss the role of mathematics in the various activities of the company.

From my experience I have concluded that visiting companies is the most important thing to do. It is the only way to learn what the world looks like from the viewpoint of applying mathematics. It is almost the only way to find case-material and items for presentations. The visits also provide an opportunity to arrange consulting contacts between universities and industry. Each mathematics department should have a "Mathematics Ambassador" shuttling between companies.

LIST OF EXAMPLES

Next I want to present a short list of examples from Finnish industries where mathematics is being used as an important agent to solve practical problems. The list is at the same time a sample of the items presented at the seminar described above.

1. Veneer cutting

The first example comes from the mechanical woodworking industry, more specifically from plywood production. Plywood is produced from thin layers of rotary-cut veneer which are glued together to form sheets. The problem is how to obtain the maximum amount of veneer from logs which may be quite irregular in shape.

The process has to be automated and one needs a strategy of centering the log between the pin and the spindle of a cutting machine. Using computer simulation a group of mathematicians compared three different centering methods. As a result a new centering strategy was found giving 9-20 % more veneer from the same amount of raw material.

The crucial point in this type of simulation is a good model for the raw material. One needs a mathematical mechanism to create plausible real looking logs in the computer memory. In this task a lot of simple mathematics was used, including statistics, random numbers, splines etc.

2. Ore prospecting

A major Finnish company working in the mining and metallurgical industry, electronics etc. has set up a group of mathematicians to probe the possibilities of using mathematical methods in ore searching.

Underground deposits containing metals differ from their surroundings in their electromagnetic parameters. As an example we can regard the conductivity σ and magnetic permeability μ as functions $\sigma(x,y,z)$ and $\mu(x,y,z)$. Now if an electromagnetic field is created in the region, it will be altered or scattered by the inhomogeneities in the ground. By observing the response on the ground it is, in principle, possible to extract information about what is concealed below.

The experimentation used is the following. A harmonic EM-transmitter is placed on the ground and the resulting total field E and B are measured over the frequency band from 0.1 Hz to 10 kHz. From this information one should try to delineate the unknown distribution $\sigma(x,y,z)$.

The group has pushed its work to two main directions.

a. Direct modelling by the "guess-and-try-again" methods.
b. Mathematical inversion methods.

Approach (a) means you change your assumption about the underground structure until (if ever) the measurements agree with theoretically calculated results. Mathematically this means solving a horrible looking partial differential- or integro-differential equation. The numerical treatment of this leads to problems in numerical analysis, integral transforms, PDE:s, integral equations, etc.

Approach (b) is a more ambitious effort to find an analytical procedure to determine the unknown σ from the measured data. From the mathematical point of view this approach leads to the so called inverse problem of the wave-equation. The form of a partial differential equation and it's solution in a certain area are known - the unknown coefficients of the equation should be determined. Experimental conditions make the problem ill-posed and the problems of numerical stability emerge as a major difficulty.

The inverse problem of the wave equation is encountered in an almost similar form in many situations like x-ray tomography, radar technics, biomagnetism etc. Motivated by this the Finnish Academy launched a research project employing a group of 4 mathematicians to study the inversion problem and its industrial applications in 1984.

Some of the numerical procedures developed during the project are already being applied in the interpretation of ore-prospecting data. The more theoretical research on the inversion problem is still at a tentative stage. Some inversion procedures applicable to simplified conditions (e.g. layers-structure) are ready for use.

3. Steel casting

The next example deals with the continuous casting of steel. Molten steel flows through an opening. The steel slab solidifies starting from the surface inwards and the solid slab then slides over supporting rollers. Cooling can be hastened by spraying water on the cooling zone.

From both the technical and metallurgical point of view there are numerous limitations in the temperature control of the surface. These limitations include

- the upper limit in the allowable length of the liquid region
- the limit in the temperature variation along the surface
- upper limits in the temperature at the rollers to avoid bulging.

The presence of both liquid and solid steel gives a special nature to the model governing the cooling process (two-phase Stefan problem).

As part of a larger joint-research project between the steel-industry and universities a group of mathematicians has been doing research on this problem. The group has devised a model together with a numerical algorithm to determine the optimal cooling in order to achieve the desired surface temperature and to meet the metallurgical criteria. In the numerical treatment of the problem both FEM and finite difference methods are used. The optimal control is calculated by the so called subgradient method.

4. Reliability modelling

The next example comes from a large electronics company. When an electronic device is sold in wholesale quantities (hundreads to thousands) the reliability of the product is a central issue. The customer wants some kind of a quarantee concerning the appearance of detects in the product.

The occurrence of defects is a stochastic phenomenon and it increases with the aging of the device. For this reason reliability specifications are defined in statistical terms and they are based on some acceptable build-up of the defect count.

The study of reliability is generally done using a stochastic mathematical model mimicking the defect behaviour of the device. For instance, long-term prediction of the defect rate is made possible by fitting an appropriate stochastic model to experimental data.

A group of mathematicians has been working for an electronics company to build a new more flexible reliability model. The company expects to get an instrument both for predicting the reliability of their products and defining the limits of acceptable defect turn-up.

The starting point of the research group has been an observation that models previously used give too pessimistic or sometimes too optimistic long-term predictions. (Models based on non-homogenous Poisson process and the so called Duane models). This error has a simple partial explanation. At the early stages of the life-cycle of a device composed of perhaps several hundred components one can see a small but recognizable peak in the defect statistics. This is caused by a small amount of originally detective components that have accidentally passed quality control. Once these have been replaced the device will resume the natural "orbit" of its life-cycle.

The models used previously neglect this fact and give erroneous results by overweighing the early part of the defect statistics. The objective of the research group has been a more adjustable model which takes this phenomenon into account.

Remarks and suggestions

Finally I want to present a few general remarks concerning the promotion of the industrial and professional use of mathematics.

1. Mathematicians are lacking a recognized professional identity. Compared to doctor, engineer, dentist, carpenter etc. there is no such thing as an industrial mathematician. The idea is quite obscure to the great public. We should develop a distinct professional profile and a self-assurance about our significance.
2. Creative development in academic curricula of mathematics is of urgent importance. What is needed is an ingenious post degree or further education program aimed at the practical skills of mathematical modelling, computing, software acquaintance and research management. This program would qualify the student to use a title "industrial mathematician", "research consultant", "risk analyst", "polymath consultant" etc.
3. The so far scanty use of mathematics in industry is to a great extent a marketing problem. Almost every field of production has created a well-organized network of distribution- and marketing infrastructure. There are different activities of
   - development
   - production
   - wholesale/retail distribution

- sales promotion
- repair service
- customer consulting etc.

Not so in the area of mathematics where the only links in this chain are the development department (= university) and the user. We should develop the "retail outlets" of mathematical knowledge and the concepts of "mathematics ambassador" and "sales consultant".

4. One of the hot words today is high-tec. There is a boom of advanced technology programs, Science Parks, Technology-centers etc. We mathematicians should hang on this development by spotlighting the evident association between advanced technology and industrial mathematics. From this perspective I see mathematics as the high technology of numbers, symbols and information processing.

LITTERATURE

1. P. Kuhno, P. Neittaanmäki, T. Tiihonen: Sorvipöllin keskittämismenetelmien vertailu tietokonesimulointia käyttäen, Paperi ja Puu 10 (1983), p. 626-629.
2. K. Aittoniemi, M.T. Hirvonen et al.: Broadhand Electromagnetic Measurements over Conductive Orebodies and their Interoretation with Heterogenous Plate Models, Acta Polytecnica Scandinavica, Applied Physics Series No 151, 1985.
3. Electromagnetic Scattering from an Inhomogeneity in a Half Space by the Method of Successive Reflections, Acta Polytecnica Scandinavica, Mathematics and Computer Science Series No 41, 1984.
4. P. Neittaanmäki: On the Control of Cooling During Continuous Casting, To appear in the Proceeding of the Fourth International Conference on Numerical Methods in Thermal Problems, July 15th - 18th, 1985, Swansea.

THE OXFORD STUDY GROUP WITH INDUSTRY, 1968-1985

S. D. Howison
Mathematical Institute
24-29 St. Giles
Oxford
England

ABSTRACT. This article is a brief description of the activities and organisation of the Oxford Study Group with Industry.

1. INTRODUCTION

There is no doubt that the astounding increase in the power and availability of computers over the last two decades has made mathematical modelling a powerful tool in industrial research. Correspondingly there has been a growing trend (exemplified by the existence of this symposium) for university mathematicians and numerical analysts to seek applications of their skills in industrial mathematics. This article is a brief outline of just one of many responses to this situation, the work of the Oxford Study Group with Industry (OSGI).

OSGI was founded in 1968 with the aim of encouraging interaction between academic and industrial research workers. The area of interest has been 'continuous' or 'physical' applied mathematics - that is, models of continuous processes which usually lead to a problem in differential equations - but this was and is because that is our main area of expertise: there is certainly also just as great scope for those interested in other fields such as statistics or operations research. The success of the venture has rested largely on the enthusiasm and team spirit of the faculty members involved, and it may be helpful to consider what the benefits to them of such an interaction are likely to be.

Firstly, there is the satisfaction of working on 'real-world' problems: the likelihood of a tangible and useful outcome to one's efforts. Secondly, there is the excitement and interest of using mathematics in a wide variety of physical situations. Thirdly, and maybe most importantly to the academic mathematician, is the stimulus to mathematical research. We have consistently found that interesting and important mathematical questions are raised by our industrial research, and the direction and style of mathematical research within the group has been considerably influenced by the industrial connection.

Conversely, there are requirements which must be met by academics who wish to interact with industry. Industrial research has a different timescale and different objectives from those in academic research, and it must be appreciated by both sides that some compromise will be necessary. Communication, too, can be a problem - what is well known to one person may be mere jargon to another - and patience is an essential lubricant of the process.

A valuable by-product has been the educational experience afforded to graduate students. A number of PhD theses have been written on industrial topics, and a one-year MSc course entitled 'Mathematical Modelling and Numerical Analysis' has been offered since 1979. About 8 students per year take the course, which involves two terms of lectures on basic and advanced modelling and numerical techniques, followed by a dissertation which usually deals with an industrial problem.

I have said nothing about cooperation with industry in numerical analysis, although this is obviously important and goes hand in hand with the mathematical interaction (indeed, many problems are brought to us for numerical advice, but turn out to need further modelling work); this is dealt with by Dr. McKee elsewhere in this issue.

Further details of the history and organisation of OSGI, together with examples of problems discussed, can be found in references 1-5.

2. THE STUDY GROUP MEETING

The focus for the activities of the Study Group is a week-long meeting held in Oxford each spring. The emphasis at this meeting is almost entirely on discussion of unsolved problems from industry, although some lectures on relevant mathematical and numerical techniques may be given. Recent meetings have been attended by several dozen people. Typically we have about ten visitors from industry, each with a problem for discussion, and some observers; the balance is made up by faculty members and students from Oxford and other universities. Indeed, the faculty contingent from outside Oxford is now so large (mathematicians from ten other universities in Britain and abroad attended in 1985) that the meeting has been renamed the Mathematical Study Group with Industry.

There has been an immense variety in the problems brought to the Study Group meetings (approximately 150 in all). They range from the very well known to the very novel, from those where the model is well understood to those where a model must be constructed from scratch. All, however, are subject to a searching scrutiny at the meeting. This variety may be illustrated by the 1985 list: in full, this reads:

- -- modelling of fluorescent dyes in fabric;
- -- explosive cliff-blasting;
- -- current leakage from a high voltage DC cable;
- -- 'pressure-swing' absorption of contaminants in a gas by a porous matrix;
- -- chemical reactions at the walls of metal vessels;
- -- collapse of tunnels in coal mines;
- -- blast furnace modelling;
- -- interaction between a drop of liquid gas and water;
- -- scattering of sound by an underwater elastic plate;
- -- numerical solutions for turbulent flow;
- -- metal-added welding.

The timetable for a typical meeting would be as follows.

Monday: The industrial visitors give a short presentation of their problems to all the participants.

Tuesday - Thursday: As much time as possible is devoted to informal discussion of all the problems. Several rooms with blackboards are available, and faculty and students gravitate towards the problems that interest them. Most problems should be looked at thoroughly from various points of view at this stage. Numerical back-up is available throughout the week, and numerical solutions are often carried out.

Friday: A summing up of progress made on each problem is given by the faculty member(s) most closely associated with it.

Reports of the progress made at the meeting are written and distributed in the weeks immediately after the meeting.

The atmosphere is as informal as possible; not the least contribution to the smooth running of the week is the relaxed social setting throughout: discussions continue late into the night over a drink, and as many solutions are proposed over dinner as anywhere else!

3. YEAR-LONG ACTIVITIES

Industrial collaboration is by no means confined to the annual meeting; in fact it continues throughout the year.

The one-year MSc course has already been mentioned. There are usually two or three students working on a PhD problem from industry, for it is common for problems brought to the Study Group meetings to have longer term aspects and implications. Some of these students are state funded, and some hold a studentship jointly funded by government and industry to work on a particular problem. There are also some postdoctoral research assistants; again, some are funded by government and some by industry.

We receive regular visits from industry, and visit industrial laboratories ourselves. Some of these visits are in connection with longer term work on problems from a Study Group meeting, while others are to discuss problems which crop up during the rest of the year.

4. CONCLUSION

I have outlined the present state of the response in one university to the opportunities available to the applied mathematician in the field of industrial mathematics. A more thorough discussion of the rewards and difficulties of university-industry interaction is unfortunately beyond the scope of this article; so, alas, is a description of any of the fascinating problems which have come our way (see references 1-5). It only remains to say again in conclusion that there are many absorbing problems to be solved, there is great intellectual stimulus and satisfaction to be found in working with industry; and that they are there for those who are prepared to go out and work hard looking for them.

REFERENCES

1. Ockendon, J.R.: 'Differential equations and industry', Math. Sci. 5 (1980) 1-12.

2. Tayler, A.B.: 'Mathematical modelling and numerical analysis', Bull. Inst. Math. App. 16 (1980) 7-10.

3. Tayler, A.B.: 'Mathematical modelling: art or science?', Math. Intell. 3 (1981) 66-69.

4. Tayler, A.B.: 'Differential equations and the real world', Bull. Aus. Math. Soc. 26 (1982) 421-443.

5. Tayler, A.B.: 'Oxford Study Groups with Industry, 1967-83', in Proceedings of the Conference on Mathematics in Industry, ed. H. Neunzert (B. G. Teubner, Stuttgart, 1984).

MATHEMATICAL MODELLING IN LIFETIME TESTING

W. Krüger, N. Neunzert
Laboratory for Technomathematics
University of Kaiserslautern

Critical safety components of mechanical systems such as cars and planes must not fail during the service life of the system. Therefore tests are needed to show that the component survives the loading posed on it during the whole service life with a given probability. How a given loading damages a component and how the component fatigues under the loading are problems for material scientists. Mathematicians however have to supply a mathematical structure or language appropriate for describing concepts like damage, lifetime etc. They have to reformulate the results of material scientists in this language in order to calculate damages created by certain loadings, to predict the lifetime of a component under such a loading and so, finally to help the engineer in testing the reliability of the system.

This paper describes the first steps in this direction. It is closely related to practical problems which arose in our cooperation with the car manufacturer AUDI. It doesn't reflect the real history of this cooperation: We first had to solve very concrete mathematical problems posed by the test engineers, shortly characterized by the words "Markov" - or "rainflow" counting - see the articles by Krüger et al in [1], [2], [3] and [4]. The solutions of these problems were practically relevant; they are now implemented in the AUDI road simulator.
But in order to go further, to be really able to improve the testing methods, one has to understand more fundamentally what is mathematically going on and what kind of mathematical ingredients are needed in order to handle the problem more generally.
First, one tries to use theories, concepts, structures at hand - as every modeller does. Are they sufficient to model the problem? Or do we need new concepts, structures and finally theories? We believe that lifetime testing is a field which needs new theories and so will give new impulses to mathematics. Certainly, the richness of theories "at hand" depends

strongly on the individual person who models (therefore modellers have to be "rich generalists"). But as far as our knowledge is concerned, the theories presently available to the modeller are not appropriate to this particular problem. Therefore, we must persue a new research program. Let's begin.

Looking at a real loading function, we realize that we should model it as a trajectory $X_t(\omega)$ of a stochastic process. Which kind of a process could that be? Is it a stationary or even an ergodic process, maybe a Gaussian, a process with continuous or differentiable paths, etc.? We have to remember that we are only interested in the lifetime connected with such a trajectory. What is the mathematical description of this connection? And is lifetime a property of the process or a property of the individual trajectory? Or, maybe, a better question: Can the stochastic process be modelled in such a way that lifetime is a property of the process?
The crucial concept here is the damage and we define it again as a process D_t, for simplicity with finitely many discrete values (damage steps 0,1,...,d) such that D_t, more precisely the probability distribution of the damage D_t at time t, depends on the trajectory $X_\tau(\omega)$ up to time $\tau = t$; shortly

$$D_t = D_t(\{X_\tau(\omega); \tau \leq t\}).$$

If the damage reaches the highest step d, we say that the component fails. Therefore, the lifetime of a component under a certain loading is given by the earliest time the damage reaches d

$$L = \sup \{t \geq 0;\ D_\tau \neq d \quad \text{für } 0 \leq \tau \leq t\}.$$

We believe that in general randomness has to enter the model twice: The loading function is random (i.e. a trajectory of a certain stochastic process) and the damage is not determined by the loading, but a random variable, whose distribution however is determined by the loading. This models somehow the fact that we know very little about the interior mechanism of damaging. Given a loading trajectory $X_\cdot(\omega)$, a certain damage trajectory $D_\cdot(\tilde{\omega})$ is realized which deterministically define the lifetime $L = L(\omega,\tilde{\omega})$.

If, for example, a given loading creates during time step Δt an increase of the damage by one step with probability q (the damage remains unaltered with probability 1-q), one calculates easily that the average lifetime is given by $\frac{b-1}{q}\,\Delta t$; here ω is assumed to be fixed and the average is taken with respect to $\tilde{\omega}$.

The problem is now again the following: Does the damage process - and therefore the distribution of the lifetime - depend only on the loading process X_t or on the trajectory $X_t(\omega)$? It is clear that the answer is strongly influenced by our device of the loading process. If it is "stochastically poor", i.e. almost deterministic as in $X_t(\omega) = A \cos(\nu t + \varphi(\omega))$, where the amplitude and the frequency are fixed and only the phase is random (for example with a uniform distribution on $[0,2\pi]$), each trajectory creates certainly the same damage (per unit time), i.e. the damage distribution at a fixed time doesn't depend on ω and the expected lifetime under these loadings is given by

$$\Gamma = \frac{2\pi}{\nu} N(A),$$

where N(A) is a certain number characterizing the mean damage created by a "cycle" of amplitude A. One may interpret therefore N(A) as the average number of cycles (i.e. periods) of amplitude A which leads to a failure. One can find in industry bench tests for shockabsorbers where the input is well modelled by this process - but then such a modelling is certainly superfluous.
On the other hand, if we generalizes this example such that $X_t(\omega) = A(\omega) \cos(\nu t + \varphi(\omega))$, where A^2 is exponentially distributed, we get a nice Gaussian process, whose trajectories are periodic with one fixed period but different amplitudes. Since the damage is greatly affected by the amplitude, the damage process will depend strongly on the individual loading trajectory. Therefore also the most important quantity with respect to lifetime testing, the mean lifetime, depends on the trajectory $X_t(\omega)$.

If one looks closer at the damage mechanisms, one will moreover come to the conclusion that if we model X_t as a Markov process, the damage process again must depend on the individual trajectory. Without going into details we just mention that this follows from the fact that the damage is influenced by the occurence of hysteresis loops in the stress/strain diagram. The damage is not only influenced by the height of a jump in the loading but depends also on the relative position in the loading trajectory where this jump occurs. This is realized in the socalled rainflow counting, which we will soon discuss. Earlier it was thought that a Markov process would be a good model for the loading, so that engineers used only "Markov counting", a method of data reduction for loading trajectories which can be used to estimate the transition probabilities of the process (see [1]).

What we would need in order to reflect the present knowledge of material scientist about damaging is a theory of "rainflow processes" as processes for which the corresponding damage process does not depend on the realization. A simulation of this process could then be used as input to a test facility (road simulator) to give estimates of the lifetime.

In order to understand this rainflow process we consider trajectories and ask for functionals which are "lifetime sensitive" in the sense that if these functionals have different values, the corresponding mean lifetime differs (one has to remember that the mean is taken with respect to the damage process, i.e. $\tilde{\omega}$).
To express it differently: Trajectories, which create the same average value for the lifetime, belong to a level set of such a functional. A set of such functionals would be "complete", if the intersection of all their level sets would determine precisely the set of all trajectories with a certain mean lifetime. To reach this completeness is again a task for material scientists, but there are some natural lifetime sensitive functionals connected with rainflow counting.

As the Fourier-decomposition in the ordinary theory of stochastic processes, the "rainflow decomposition" of a trajectory should play an essential role in the new theory. We explain it with an example (see [1], [2] for details):

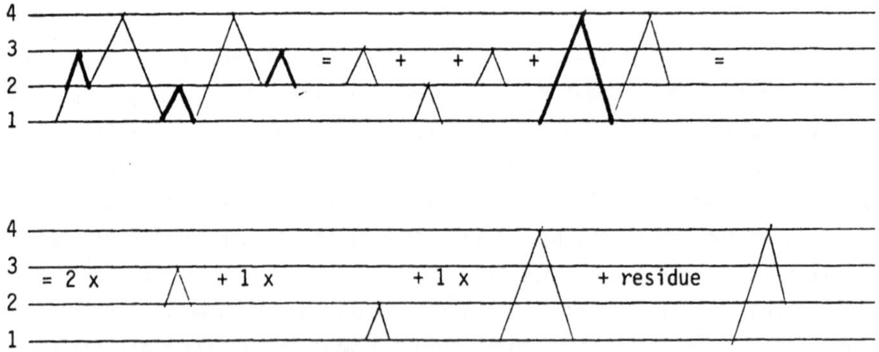

It is a decomposition into "intermediate cycles" and the residue where the others can be imbedded. If we define as before N(A) as the average number of cycles of amplitude A which leads to a failure, then the mean value of the damage created by a trajectory which consists during the inverval [0,T] of a series of n_i cycles of amplitude A_i, $i = 1,\dots,k$, may be approximated by

$$\overline{D}(X_.) := b\cdot \sum_{i=1}^{k} \frac{n_i}{N(A_i)}$$

(where b is the amount of damage that leads to failure). Therefore, if $\overline{D} = b$, i.e. if

$$f(X_.) := \sum_{i=1}^{k} \frac{n_i}{N(A_i)} = 1$$

we expect a failure during the time [0,T]. f is certainly lifetime sensitive.

In our example we have (forgetting the residue)

$$f(X.) = \frac{3}{N(1)} + \frac{1}{N(3)} .$$

Typically, $N(A) = A^{\kappa}$, where κ is around 5 for typical materials.

If we count the number of cycles $i \to j \to i$ in a symmetric matrix F (this is the rainflow matrix), our example gives

$$\begin{pmatrix} 0 & 1 & 0 & 1 \\ 1 & 0 & 2 & 0 \\ 0 & 2 & 0 & 0 \\ 1 & 0 & 0 & 0 \end{pmatrix} \quad + \text{ residue } (1,4,2)$$

f may be interpreted as a matrix norm

$$f\,(X_.) = \|F\| := \sum_{i,j=1}^{N} f_{ij}\,|i-j|^{-\kappa}$$

One has to realize that this functional does not depend on the behaviour of the trajectory "in time". For example it does not depend on the velocities of the jumps. Moreover it counts only the amplitudes of the cycles and not the levels at which these cycles occur. If one wants to take into account at least the second effect, one gets immediately N^2 lifetime sensitive functionals, namely all the coefficients f_{ij} in the rainflow matrix F. Different rainflow matrices mean different damage processes! Not all these functionals (included the values in the residue) are independent, there are some conditions a pair (matrix, residue) = (F,R) must fulfill in order to be the rainflow data of a trajectory: For example a cycle must "fit" into the residue. A full characterization of rainflow data is in principle simple but a little bit complicated to write out (see [1] and [2]).

The concepts of damage processes and of lifetime sensitive functionals and the practical use of these concepts in test facilities lead to new mathematical structures and problems.
An example is the practical task of getting information about the reliability of car components in long distance tests by observations on rather short distances. One has to "compose" pieces of loading functions, to "extrapolate" from 200 km to 300 000 km. This defines something like a vectorspace structure on the set of trajectories, where addition is not the usual superposition and multiplication by a scalar is also not the usual one.
If we denote by X_T a trajectory defined for $0 \leq t \leq T$, then we define

$$X_{T_1} \oplus X_{T_2} := X_{T_1+T_2} \quad \text{with} \quad X_{T_1+T_2}(t) := \begin{cases} X_{T_1}(t) & \text{if } t \leq T_1 \\ X_{T_2}(t-T_1) & \text{if } t > T_1 \end{cases} .$$

Are the rainflow data linear with respect to this operation $\oplus$? Not really, but almost; $F(X_{T_1+T_2})$ and $R(X_{T_1+T_2})$ can be calculated as follows:

Consider $R(X_{T_1}) \oplus R(X_{T_2})$ as a loading function with rainflow data $F(R(X_{T_1}) \oplus R(X_{T_2}))$, $R(\ldots)$; then $(F(X_{T_1}) + F(X_{T_2}) + F(R(X_{T_1}) \oplus R(X_{T_2})), R(\ldots))$ $= (F(X_{T_1+T_2}), R(X_{T_1+T_2}))$ is generally not true, since the left hand side may fail to form a rainflow pair. But minor changes in F, which consist in substituting cycles $i \to j \to i$ by cycles $j \to i \to j$ lead to $F(X_{T_1+T_2})$.

Here an example:

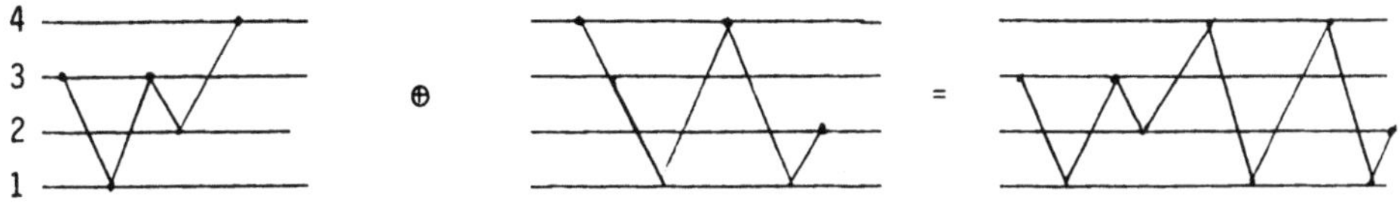

$a_{32} = 1,\ R = (3,1,4)$ $\qquad$ $a_{14} = 1,\ R = (4,1,2)$ $\qquad$ $a_{32} = a_{41} = 1$, $R = (3,1,4,1,2)$

$R(X_{T_1}) \oplus R(X_{T_2}) = (3,1,4,2,2)$, so that
$F(R(X_{T_1}) \oplus R(X_{T_2})) = 0$ and
$F(X_{T_1}) + F(X_{T_2}) + F(R(X_{T_1}) \oplus R(X_{T_2}))$ is
given by $a_{32} = a_{14} = 1$. Changing $1 \to 4 \to 1$ in
$4 \to 1 \to 4$ one gets $a_{32} = a_{41} = 1,\ R = (3,1,4,1,2)$ = "

A different addition + leads also to a different scalar multiplication. However, $nX_T := X_{nT} = X_T \oplus \ldots \oplus X_T$ would be only mathematically, not physically reasonable. If one drives the car over 1 km and never gets a very strong impact (such that $a_{ij} = 0$ for $|i-j|$ sufficiently large), the same is not necessarily true when one drives over 300.000 km, but the elements of X_{nT} always have the property of being zero sufficiently far away from the diagonal.
Instead of defining the extrapolation to times like nT for large n by nX_T it is better to consider the rainflow matrix from a statistical point of view, i.e. consider the coefficients as numbers of events having a two-dimensional distribution, whose parameters must be fitted to the experimental data (see [2]). Defining X_T in this way destroy the vectorspace structure, so that one may believe that the definition of $\oplus$ is not the appropriate one - certainly there is room for improved models.

One may also normalize the coefficients of the rainflow matrix such that they can be considered as relative frequencies approximating the probability of cycles. Interpreting stochastic matrices in this way (and not as Markov matrices with transition probabilities) should lead to a new class of stochastic processes, perhaps "rainflow processes", appropriate for modelling damage processes where hysterese loops (cycles) are important. But this is another

new task for the mathematicians!
The practical use of rainflow data has created another mathematical task; the reconstruction. The idea is as follows: A real car is driven over a test road so that a real loading is created. The rainflow data of this loading are stored on-line. The car returns with a rainflow matrix and a residue. Now one wants to use a test rig, which creates an artificial loading of the car similar (for the point of view of lifetime) to the real one. The mathematical problem of reconstruction is: Given rainflow data; construct a loading function which corresponds to these data. In solving this problem one has to realize that the rainflow data do not determine the loading function (not even the sequence of maxima and minima of the loading, which is the only important feature). However, since we have finitely many states there are only finitely many solutions of the reconstruction problem. We have to construct a random-path-generator, which selects each of these solutions with the same probability. This problem is solved and the solution has been used in a practical situation [1].

Let us finish by looking to future problems posed by test facilities. Up till now the loading was one-dimensional - in reality the loading is at least three-dimensional. The three components of the loading function are certainly correlated as functions of the time - if the car drives over a pothole, the car component in question will be loaded in every direction, approximately at the same time. The reconstruction algorithm now available forgets time and is therefore not able to reflect these correlations. It could fail in predicting the lifetime. Moreover, rainflow data are lifetime determining only if the road is near to white noise. If one has a periodic behaviour of the road, for example in a cobbled pavement, certain components of the car may oscillate in resonance with the pavement. The frequency (out of consideration in rainflow counting) becomes important in this way.

These two arguments (and more) make the consideration of time in reconstruction necessary. Fourier analysis is not easily combined with rainflow procedu - it relies on the normal vectorspace structure of function spaces, and not on $\oplus$.
A first step is done in [5]: Empirical data led to the idea that the duration of a cycle $i \to j \to i$ is Γ-distributed with parameters depending on (i,j). One gets an additional matrix describing these parameters and another set of

lifetime sensitive functionals. The Γ-distributions can be used for the reconstruction which then reflects somehow the time behaviour of the original loading. Whether it also reflects the frequency content of this function is an open and interesting question..

Let us finish: We are convinced that damage processes, lifetime sensitive functionals, rainflow processes are fruitful concepts - for the material sciences and for mathematics as well. Today they are not yet a part of a nice theory. A research program in this area promises many theoretical and practical rewards. And this is the best a cooperation with industry can provide.

Literature

[1] W. Krüger, M. Scheutzow u. A. Beste, J. Petersen: Markov- und Rainflow-Rekonstruktionen stochastischer Beanspruchungszeitfunktionen; VDI-Fortschrittsbericht, Reihe 18, Nr. 22, 1985

[2] W. Krüger, J. Petersen: Simulation und Extrapolation von Rainflow-Matrizen; Berichte der AGTM, Nr. 8, Kaiserslautern 1985

[3] W. Krüger: Simulation stochastischer Prozesse für den Lebensdauernachweis beim Pkw; in: H. Neunzert (ed.) Proceedings of the Workshop "The Road-Vehicle-System and Related Mathematics", Teubner 1985

[4] J. Petersen: Evaluating Durability of Passenger Car Components: Measurement-Calculation-Simulation; in: H. Neunzert (ed.) Proceedings of the Workshop "The Road-Vehicle-System and Related Mathematics", Teubner 1985

[5] M. Keul: Mathematische Modelle für das Zeitverhalten stochastischer Beanspruchungszeitfunktionen, Berichte der AGTM, Nr. 12, Kaiserslautern 1986

INTERFACES BETWEEN OPERATIONS RESEARCH AND COMPUTER SCIENCE

J.K. Lenstra
Centre for Mathematics and Computer Science (CWI), Amsterdam

The subject of this lecture is the relation between two disciplines, both of which pertain to the use of mathematics in industry. Operations research emerged out of optimization questions in warfare logistics in the 1940's and quickly established itself as one of the cornerstones of industrial mathematics. Computer science and engineering is a younger discipline, which pervades all activities relating to the processing and manipulation of quantitative information at an ever increasing rate.

I cannot possibly give a complete overview of all interfaces between operations research and computer science. The time is too short and my background is too limited. So you should expect a biased view of some interfaces.

Let me give a short preview. I will make only a few remarks on the impact of operations research on computer science and concentrate on the influence in the other direction. Another bias is the emphasis on the combinatorial side of operations research rather than on the nonlinear or stochastic ones. I will go through issues of computation, complexity and analysis of algorithms, and then mention three new tools: randomization, parallelism (both quite briefly), and interaction. A substantial part of my lecture will be devoted to interactive computing or, more precisely, to the integration of interaction and algorithmics in what is nowadays called "decision support systems". I will finish with some remarks on the subject of expert systems.

Operations research vs. computer science

The basic claim which I would like to make is that operations research and computer science cannot exist without each other. More specifically, the evolution of either area to a discipline in itself has only been possible due to the use of concepts, results and techniques developed in the other area.

As I said before, I will concentrate on the impact of computer science on operations research. That is not to say that the influence in the other direction is less important. It is hard to resist the temptation to spend the entire hour on this subject. Many applications of operations research occur in the design, analysis and control of computing devices. This has been true since the very beginning of automated computing. The three examples I will give are the more topical applications of the 1980's.

First, the *layout of integrated circuits*. Locating several given components on a chip and routing the required connections between them in such a way that the total area is minimized, is a highly complex problem. It is, in fact, a combinatorial optimization problem, and many ideas from location, routing and scheduling theory could contribute to its solution. It is my impression that this interface is at a very early stage of exploration.

On a larger scale, the *performance of computer systems* is a subject of increasing interest. From an operations research point of view, it calls for the analysis of the behavior of communication networks and thus belongs to the application area of queueing theory. This interface is in rapid development.

In the third place, the *design and control of distributed systems* leads to many algorithmic problems. Consider, for example, a network of processors that cooperate to perform a certain task. Each processor has its own memory. Which data must be stored where so as to achieve a reasonable balance between storage and access cost? This is just an example of a question in distributed computing to which operations research could contribute.

It appears that computer science needs operations research. In that sense, it has already a substantial impact by providing a wealth of challenging decision problems.

The influence that I will emphasize, however, is more fundamental. Without the achievements of computer science, the practical application of operations research methods as well as our theoretical understanding of their behavior would have been very limited.

Computations, efficiency, and complexity

Operations research needs computer science. This was already evident at the origin of our discipline. The whole idea of solving large scale decision problems by iterative techniques was only a feasible idea in view of the existence, or the prospect of developing, automated equipment for performing massive amounts of computation. This applies to the work on linear programming and network optimization that started in the late 1940's. It is also true for most of the subsequent methodological developments, such as cutting plane methods, dynamic programming, branch and bound, and Markov programming.

So far, the influence of computer science was basically restricted to providing the machinery needed. During the 1960's, the impact deepened and acquired a component that was more algorithmic in nature. Operations researchers became interested in the efficiency of computer algorithms. In order to reduce the time and space requirements of their implementations, they had to use sophisticated data structures. With this, operations research started to rely on the art of computer programming.

The advent of computational complexity theory in the beginning of the 1970's can, in retrospect, be seen as a natural culmination of this development. At the time, it was a surprising and exciting event. Complexity theory studies the inherent limitations to the efficiency of algorithms. It provides a simple tool to distinguish between the tractable problems and the intractable, or probably intractable, ones. This distinction is now common practice in combinatorial operations research. For the easy problems, a fast optimization algorithm is available. For the hard problems, one has to choose: if an optimal solution is required, then one must settle for some tedious form of enumeration; if speed is desired, then one must be satisfied with an approximate solution. Linear programming is an example of an easy problem; integer programming is probably hard.

In what follows, I shall first take a closer look at the concepts of complexity and then return to the study of algorithms.

A closer look at complexity

You all know what a graph is: a collection of nodes and a collection of edges, each of which links two nodes together. The graph in Figure 1 is connected, because you can get from each node to any other. The graph in Figure 2 is disconnected. The graph in Figure 3 is Hamiltonian, since there is a cycle which visits each node exactly once; such a Hamiltonian cycle is indicated by wiggly lines. The graph in Figure 4 is not Hamiltonian; you might want to prove this.

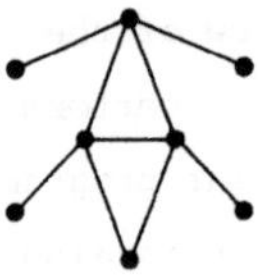

Figure 1 A connected graph.

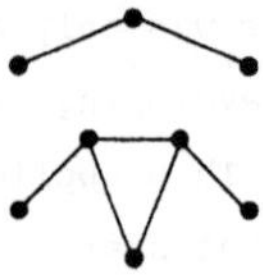

Figure 2 A disconnected graph.

Figure 3 A Hamiltonian graph.

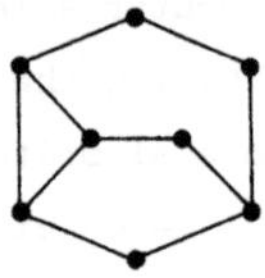

Figure 4 A non-Hamiltonian graph.

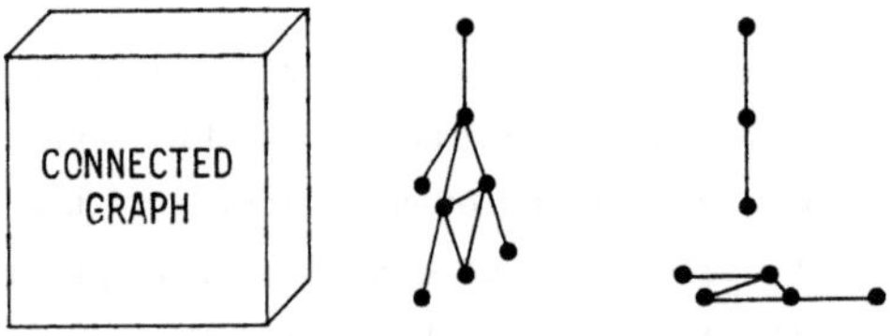

(a) ? (b) Right. (c) Wrong.

Figure 5 Buying a connected graph.

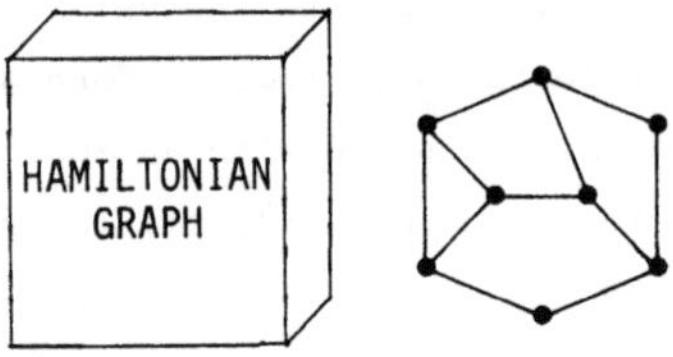

(a) ? (b) ??

Figure 6 Buying a Hamiltonian graph.

Suppose that you are attending a conference and you want to buy a present for the people at home. You go to a graph store and ask for a connected graph. The shopkeeper puts a box on the counter (Figure 5(a)). You want to check this, open the box, and take out the graph. When it sticks together, it is connected (Figure 5(b)); when it falls apart, it is not (Figure 5(c)). The point I want to make here is that you can easily test a graph for connectivity by yourself; it takes an amount of time proportional to the number of edges.

Now you want to buy something special: a Hamiltonian graph. Again, there is a box (Figure 6(a)). You open it - but now you may find yourself in trouble (Figure 6(b)): there is no fast method available which can test any given graph for Hamiltonicity. Trial and error may work, but it does not have to. However, the shopkeeper can easily convince you, namely by pointing out a Hamiltonian cycle as in Figure 3; this takes an amount of time proportional to the number of nodes.

This is exactly the difference between the problem classes P and NP. Both classes contain only decision problems, which require a yes/no answer; I will return to optimization problems shortly. P contains all those problems for which one can easily come up with the correct answer. NP contains all those problems for which one can easily be convinced of the correctness of the yes answer by checking a given structure: a Hamiltonian cycle in the example, a "certificate" in terms of complexity theory.

These definitions only make sense if the notion of "easiness" is formalized. A computation is easy if its running time is bounded by a *polynomial* function of the size of the problem under consideration. For a graph on n nodes, checking all nodes or all edges takes time polynomial in n, but generating all permutations of the node set in the hope of finding a Hamiltonian cycle is superexponential.

What are the virtues of an algorithm when it runs in polynomial time? First of all, its robustness. An algorithm that is polynomial on one machine is polynomial on any other reasonable type of machine, including theoretical models and commercial computers (but excluding parallel machines). Secondly, its asymptotic behavior. Any polynomial function in n is ultimately, when n is large enough, smaller than any superpolynomial function. In the third place, its practical efficiency. Polynomial algorithms tend to work well in practice. Some polynomial algorithms are pretty bad, but it seems to be the case that once a problem has been shown to belong to P, a truly efficient method is found sooner or later. Finally, polynomiality allows us to come to grips with computational complexity in a theoretical sense. It serves to explain why some problems appear to be harder than others. More generally, it has proven to be a fundamental concept in the broad area of computational mathematics.

Any problem in P also belongs to NP, so P is a subclass of NP. I have indicated that the connectivity problem is a member of P and that Hamiltonicity is in NP. If it could be shown that Hamiltonicity is outside P, then the problem would have no solution in polynomial time and one would justifiably call it "hard". Such a proof seems to be beyond the reach of present-day mathematics. However, we can do slightly less. It can be shown that the Hamiltonicity problem is a generalization of any other problem in NP. Hamiltonicity is *NP-complete*, i.e., it is representative of the entire class NP. If Hamiltonicity would belong to P, then all other problems in NP would be easy as well and P would be equal to NP. No one believes this to be true, for the simple reason that NP seems to be so much richer than P. It follows that the Hamiltonicity problem is unlikely to be easy and therefore "probably hard". See Figure 7.

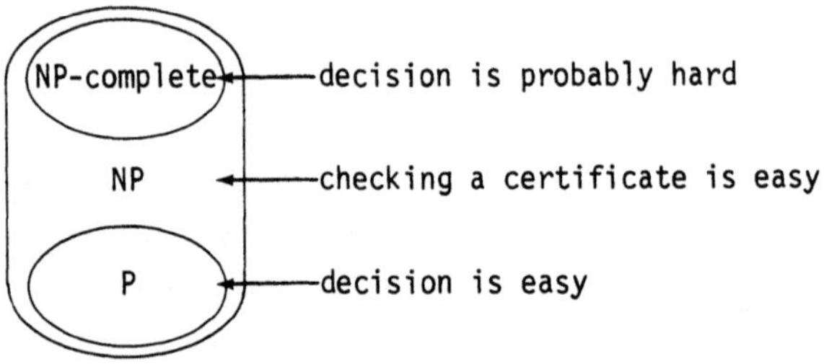

Figure 7 A likely map of NP.

Next to Hamiltonicity, many other combinatorial decision problems have been shown to be NP-complete. I have not told you how results of this type are obtained. That is of secondary importance here; suffice it to say that it is conceptually a simple affair, although it can be technically very intricate.

What is the use of all this for operations research? Complexity theory deals with yes/no problems, in operations research we have optimization problems. If a problem has a polynomial optimization algorithm, then it is said to be easy (or well solved, or tractable). If a problem is at least as hard as some NP-complete problem, then it is said to be NP-hard. This should not be the last word on the problem, but the first. It tells you that you cannot expect to find a guaranteed optimum in worst case polynomial time. You have to give in on either speed or solution quality.

Analysis of algorithms

Complexity theory has given a new impetus to the analysis of algorithms. In studying the behavior of an algorithm, you have to distinguish between its efficiency and its effectivity. Efficiency measures the resources you need: time, space, or (in case of parallel computing) the number of processors. Effectivity measures the extent to which you reach your goal. As the goal is usually an optimum solution, effectivity stands for solution quality or, more precisely, for the absolute or relative difference between the solution value obtained and the optimum value. It is the principal performance criterion for approximation algorithms.

The traditional way to analyze the behavior of an algorithm is the *empirical* one. You actually run one or more algorithms on one or more computers using one or more test problems and tabulate the results. There are some difficulties: Are the implementations of the algorithms of the same quality? How do the computers compare? Do the test problems form a fair collection? And, finally, how do you statistically validate the experimental results? In spite of this, empirical analysis is widely applied and generally very useful. However, it is not the only resort. Two new approaches have emerged, of a more theoretical nature: worst case analysis and probabilistic analysis.

The purpose of a *worst case* analysis is to provide guarantees on the performance of an algorithm. Complexity theory is basically concerned with this type of analysis. Performance guarantees are solid but may be pessimistic, since the isolated difficult problem instance has to be accounted for. The simplex method for linear programming, for example, requires exponential time on a class of artificial instances, but the method performs quite satisfactorily in practice. It is reassuring to know that a list schedule of jobs on identical parallel machines is never longer than twice the optimal schedule, but it is usually much closer.

Worst case analysis can give a misleading picture of the typical case. Thus the ultimate explanation of why algorithms behave as they do must be of a *probabilistic* nature.

A probabilistic analysis requires first of all the specification of a probability distribution over the set of all problem instances. Several random graph models have been well studied, but for many other combinatorial structures the choice of a reasonable probability model is far less obvious. Moreover, the technical difficulties encountered in a probabilistic analysis are formidable. The main reasons for this are the special structure of problem instances and solutions, as well as the interdependence between the various steps of an algorithm. What happens at a node of a search tree, for example, depends highly on what has happened at its predecessors, and no real way has been found around the resulting mathematical obstacles.

Nevertheless, progress has been made on various fronts. One of these is probabilistic efficiency analysis, an approach that is now standard for the basic algorithms in computer science. In operations research, a substantial amount of work has been done to explain the success of the simplex method. A great challenge here is to give rigorous proofs of the polynomial expected running time of various enumerative methods, in order to confirm informal analyses or empirical evidence. Secondly, there is the area of probabilistic effectivity analysis. The empirical behavior of approximation algorithms suggests that the worst case is seldom met in practice, but theoretical verification remains very difficult. Most research of this type is actually based on probabilistic value analysis, the third and perhaps most surprising area. Many hard optimization problems, notably those with a geometric structure such as routing and location problems in the plane, allow a simple probabilistic description of their optimum solution value in terms of the problem parameters. The shining example here is the planar traveling salesman problem: the length of a shortest tour through n cities, uniformly distributed over a circle of area 1, is almost surely equal to $\beta\sqrt{n}$, where β is a constant that can be estimated numerically.

Three new tools

Computer science has enabled us to perform large scale computations. It has taught us to achieve efficiency and also to accept the limits to efficiency. It has indicated how to attempt a formal analysis of algorithmic behavior.

In the last few years, it has given us three new tools: *randomization*, *parallelism*, and *interaction*. Randomization and parallelism are the most important new algorithmic concepts in theoretical computer science. Parallelism and interaction have become relevant topics due to the availability of new computer architectures. And interaction and randomization represent new modes of approximation: a randomized algorithm tosses a coin at certain points in order to decide how to proceed; an interactive method is not even completely algorithmic but relies on man-machine interaction. I will make a few brief remarks on two of these subjects and spend the rest of my time on the third one: interaction.

In a randomized algorithm, the stochasticity is inside the algorithm and not in the problem instances, as in the case of probabilistic analysis. A randomized algorithm makes guesses along the way and hence mistakes, but with a bounded probability. One of the earliest examples is Rabin's primality test. It runs in polynomial time and gives moral, although no absolute, certainty about the primality of the input number. The most topical example in operations research is the principle of simulated annealing. This is a technique for iterative improvement, based on neighborhood search, which accepts deteriorations with a small and decreasing probability in the hope of avoiding bad local optima and getting settled in the global optimum. The investigation of the randomization principle in operations research is still at a very early stage. Much can be expected along these lines in the near future.

Parallel algorithms are designed to be executed on a collection of processors that operate in parallel and communicate with each other. For certain formal models of parallel computation, algorithms have been designed for the basic problems in computer science as well as for many optimization problems. The entire theoretical approach to the complexity of problems and the efficiency and effectivity of algorithms is being extended to this area. At the more practical side, many types of parallel and pseudoparallel architectures are now becoming available, from mainframes for vectorized computations to networks of micros for distributed computations. This leads to a broad range of research questions. For computer scientists: How do the results for formal models of parallel computing translate to realistic models? For operations researchers: How do we solve problem X on architecture Y? Is it a good idea anyhow? And for everyone: Are there one or two models that will be accepted as standards for the future?

CAR: Computer Aided Routing

The vehicle I will use to discuss interactive computing is CAR, a system for computer aided routing that we have been developing in Amsterdam over the past two years.

Let me start by describing the practical decision situation in which the first release of CAR will be installed. It concerns the operational distribution planning for the hanging garment division of Van Gend & Loos, the largest Dutch road transportation firm. They have one central depot, a number of vehicles, and each day a collection of customers. Each vehicle has its own capacity. Each customer has a demand, a time window (i.e., a time interval in which he must be visited), and a priority (indicating if the visit may be postponed until tomorrow or not). The time windows appear to be increasingly restrictive, especially for garment shops located in urban pedestrian areas. In addition to this, some customers may have a supply rather than a demand, and the combination of supply and demand may lead to precedence constraints between customers. The purpose of our involvement is to help in developing a tool for improved planning. Improved in terms of lower costs, better service to the customers, and a more even work load for the drivers.

An important requirement, which we imposed ourselves, is the functional flexibility of the system. It should enable the planner to perform the traditional planning more efficiently on the one hand, and it should be able to construct a complete plan by itself on the other hand. In different words, it should assist as an automatic scratch pad and advise as an automatic pilot.

Another requirement is of major concern to the information engineers of Van Gend & Loos. The system should be implemented on a small configuration at the loading site. But it should also be integrated into the administrative organization of the firm. In the current situation, the planning is done by hand and the registration is done at a later stage on the firm's mainframe. In the new situation, the information of each order will be fed into a microcomputer at a much earlier stage. Both machines need to be linked, and Van Gend & Loos has to seriously reconsider the structure and control of its information flows.

The mathematical model motivated by this practical decision situation is a fairly typical vehicle routing problem. We have to find a tour for each vehicle, starting and finishing at the depot and collectively visiting all customers, such that three conditions are satisfied. First, the total load allocated to any vehicle should not exceed its capacity at any point in time; this is the *clustering* aspect. Secondly, the departure time at any customer should fall inside his time window (early arrivals are allowed but lead to waiting time); this is the *scheduling* aspect. Thirdly, the total travel time should be minimized; this is the *routing* aspect.

The overall problem is very hard and you cannot hope to solve it to optimality for realistic problem sizes. In fact, each of the three aspects I mentioned represents a problem which is NP-hard, independently of the other two aspects. We have chosen to follow an approximative two-phase approach suggested by Fisher and Jaikumar: in the first phase, we cluster the customers into

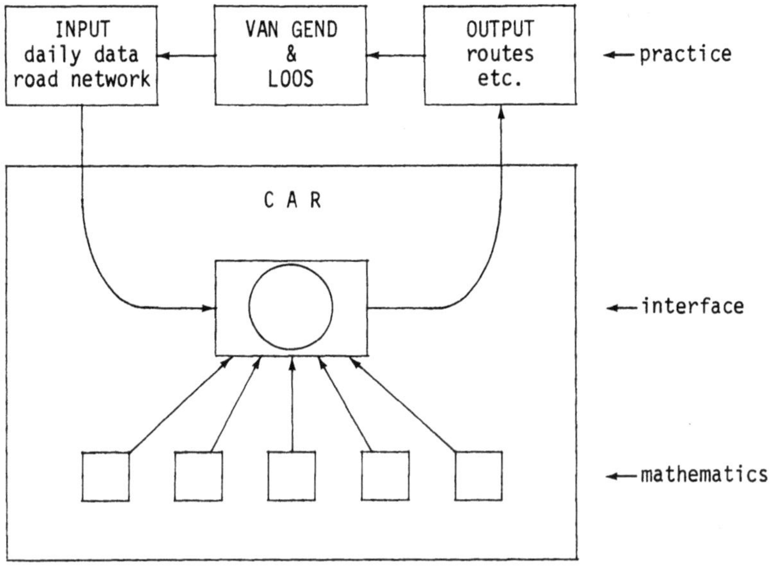

Figure 8 CAR.

vehicles by solving a generalized assignment problem; in the second phase, we route each vehicle through its customers by solving a traveling salesman problem. This does not yet take care of the scheduling aspect. Time windows are a relatively neglected complication in vehicle routing theory. We have designed routing algorithms that take account of time windows as efficiently as one could hope for, but their incorporation in the clustering phase remains a challenge.

After describing practice and sketching the mathematics inspired by it, I now want to discuss the implementation of the entire system. See Figure 8 and note that there are three levels. The top level stands for *practice*, consisting of the input to CAR, its output, and the rest of the outer world. The bottom level represents *mathematics*, a collection of algorithmic modules for solving several types of well-defined subproblems: clustering, routing, routing subject to time windows, and so on. The modular setup facilitates an extension of the system to a broader range of practical situations. Extensions that we would like to attack in the near future relate to multiple depots and heterogeneous commodities. The middle level is the core of the system: the *interface* between practice and mathematics where the man-machine interaction takes place. Its implementation heavily relies on advanced information technology for the graphical display of data and solutions in a variety of ways. Some technical data: CAR is written in the C programming language and uses a C implementation of the Graphical Kernel System; implementations are available on an IBM PC/AT with an IBM Professional Graphics Display and on an IBM 5160 (PC/RT) with an IBM 5085 Graphics Display.

I should emphasize again that the solution approach followed by CAR is not purely algorithmic but that the man-machine interaction is an essential feature. Interaction has a threefold advantage in that it adds to effectivity, efficiency and acceptability. First, the cooperation between man and machine leads to better solutions. The machine cannot be beaten in solving well-defined detailed problems. The human planner is superior in judging fuzzy situations, in recognizing global patterns, and in observing *ad hoc* constraints which do not form part of the underlying models. Secondly, these better solutions are obtained faster, because interaction allows for flexibility in manipulating data and in selecting solutions. Finally, an interactive system is more readily accepted. The human planner is not replaced by a black box but gets a versatile tool.

Decision support systems

At the CWI we are involved in the development of other interactive planning systems in which the same design philosophy is applied. One of these concerns the operational production planning for assembly lines in the Dutch clothing industry. Rather than going into any detail here, I would like to discuss the combination of interaction and algorithmics in more general terms.

In the few formal descriptions of decision support systems which I have seen, whether they were logical, technical or functional, there was always an implicit three-level structure, which is made explicit in Figure 9. The core of the system is the man-machine interaction, the dialogue between user and computer. On the practical level, the input consists of all kinds of data and also "scenarios", i.e., solutions and strategies proposed to the system. The output is, of course, decision support. And the communication with other systems is a very important subject but it falls outside the decision support system. On the mathematical level, there is a collection of quantitative models and methods.

The system must be able to perform a broad range of functions. This range should include the assisting role of an automatic scratch pad and the advisory

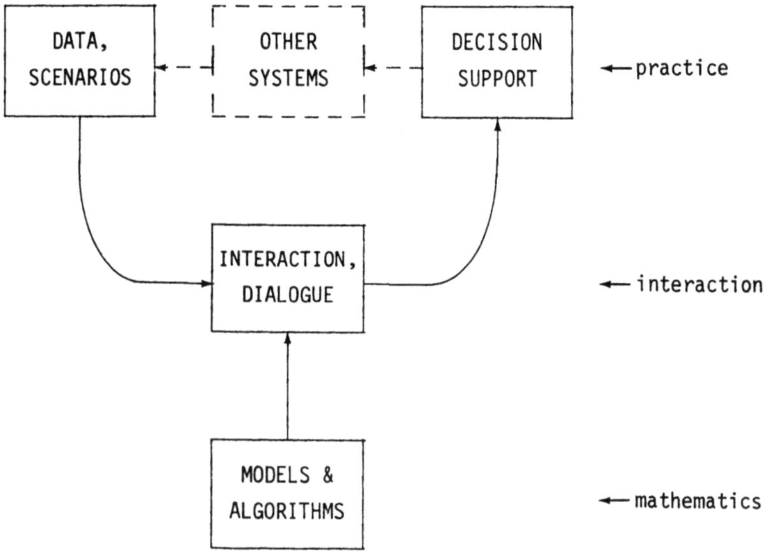

Figure 9 Decision support system.

role of an automatic pilot, which I already discussed in the context of CAR. In the first role, the interaction is essential and the computations are usually restricted to "what-if analyses", i.e., evaluations of given scenarios. In the second role, the algorithmics is essential; it should be sufficiently powerful to propose solutions of a reasonable quality.

I do not want to attempt to give anything like a definition of what a decision support system is. I do want to claim, however, that the ability to play both roles is a necessary functional condition for a system in order to qualify as such. The roles are at the extremes of the spectrum, and there is much inbetween. The aim to integrate the functions of assistant and advisor implies the need to combine interaction and algorithmics. This makes interactive decision support one of the great challenges on the interface between operations research and computer science.

Let me briefly compare traditional operations research with this concept of a decision support system. On the practical level, decision making, sometimes in terms of commands, is replaced by decision support, in terms of suggestions. On the level of information technology, the classical black box becomes transparent. On the mathematical level, the algorithms are no longer at the core of the system. They are less visible to the user and may seem to be of secondary importance. However, for the investigator who is primarily motivated by the mathematics of operations research, a decision support system is like the wooden horse of Troy, enabling him to disguise his models and methods in an attractive fashion. Information technology provides facilities for manipulating information, but these facilities only pertain to the form. The practical situation in question has to give substance to the information, and some sort of mathematical abstraction is needed to make the manipulation meaningful.

Expert systems

I would like to spend the last few minutes on expert systems. Please do not ask me what an expert system is. As the term suggests, it should encapsulate human expertise, in terms of knowledge and inference power, for a certain application area. But it remains a very vague notion, at least in operations research. The expertise of today may be common knowledge tomorrow. Sometimes a straightforward collection of heuristic rules is called an expert system because it outperforms experts. That is fine, but it deprives the notion of its contents.

I will sketch one type of system which, in my opinion, deserves the name. It receives as input the description of a problem type or situation (rather than the data of a specific problem instance). It gives as output a mathematical model and a suggestion of a suitable algorithmic approach (rather than a specific numerical solution). Input and output are concepts of a higher order than we are used to in traditional operations research or interactive decision support. And the transformation can only be made on the basis of a formal representation of operations research expertise.

Back in 1975, we built a system of this type. It is called MSPCLASS and handles a class of 4,536 deterministic machine scheduling problems. As any expert system, it has a knowledge base and an inference engine. The knowledge base consists of a subclass of known easy problems, a subclass of known NP-hard problems, and a partial order on the entire class. The partial order is denoted by an arrow: →; X→Y means that problem Y is at least as hard as problem X. The inference engine applies four rules:

- if X→Y and Y is easy, then X is easy;
- if X→Y and X is NP-hard, then Y is NP-hard;

- if X is easy and there is no easy Y with $X \rightarrow Y$, then X is maximally easy;
- if Y is NP-hard and there is no NP-hard X with $X \rightarrow Y$, then Y is minimally NP-hard.

MSPCLASS registrates our current knowledge by partitioning the entire class into three subclasses of easy, NP-hard and open problems. It also determines the borderlines between the subclasses by identifying the maximal and minimal problems in each subclass. This feature turned out to be helpful in suggesting future research.

In a technical sense, there is nothing sophisticated about MSPCLASS. All we need are elementary operations on the product of seven directed graphs. In a functional sense, MSPCLASS represents the bottom of the type of system I am discussing. The algorithmic suggestion implied by the complexity classification is certainly a very global one. However, in a conceptual sense, MSPCLASS fully qualifies, although we did not realize this at the time.

One of the things we intend to do in the near future is to build a system of this type for a broader input class of more practical relevance, the routing and scheduling of vehicles and crews, and also with a broader range of output statements than the bare distinction between easy and NP-hard. The system might consist of two phases: one phase in which a question-and-answer game transforms a practical decision situation into a rough model, and a second phase in which true expertise is applied to strip the rough model to a tractable model. This might lead to a third phase in which a decision support system is selected and applied.

It is in the development of such systems where the needs from practice and the possibilities offered by theory, from complexity theory to mathematical programming, are meeting.

Acknowledgement

The work which I reported on decision support and expert systems is carried out by the combinatorial optimization group at the Centre for Mathematics and Computer Science (CWI) in Amsterdam. The views which I expressed on these subjects are based on discussions with members of the group. I am pleased to acknowledge the contributions by J.M. Anthonisse, G.A.P. Kindervater, B.J. Lageweg, and M.W.P. Savelsbergh.

Bibliographical notes

A volume of annotated bibliographies [O'hEigeartaigh *et al.*, 1985] provides classified reviews of the recent literature in four interface areas, namely complexity [Papadimitriou, 1985], probabilistic analysis [Karp *et al.*, 1985], randomization [Maffioli *et al.*, 1985], and parallelism [Kindervater & Lenstra, 1985]. A tutorial introduction to the latter subject is given by Kindervater & Lenstra [1986]. For the Fisher-Jaikumar algorithm and much more on vehicle routing, see the annotated bibliography and the survey by Christofides [1985a, 1985b]. MSPCLASS is described by Lageweg *et al.* [1982].

References

N. Christofides (1985a). Vehicle routing. [O'hEigeartaigh *et al.*, 1985], 148-163.

N. Christofides (1985b). Vehicle routing. [Lawler *et al.*, 1985], 431-448.

R.M. Karp, J.K. Lenstra, C.J.H. McDiarmid, A.H.G. Rinnooy Kan (1985). Probabilistic analysis. [O'hEigeartaigh *et al.*, 1985], 52-88.

G.A.P. Kindervater, J.K. Lenstra (1985). Parallel algorithms. [O'hEigeartaigh *et al.*, 1985], 106-128.

G.A.P. Kindervater, J.K. Lenstra (1986). An introduction to parallelism in combinatorial optimization. *Discrete Appl. Math. 14*, 135-156.

B.J. Lageweg, J.K. Lenstra, E.L. Lawler, A.H.G. Rinnooy Kan (1982). Computer-aided complexity classification of combinatorial problems. *Comm. ACM 25*, 817-822.

E.L. Lawler, J.K. Lenstra, A.H.G. Rinnooy Kan, D.B. Shmoys (eds.) (1985). *The Traveling Salesman Problem: A Guided Tour of Combinatorial Optimization*, Wiley, Chichester.

F. Maffioli, M.G. Speranza, C. Vercellis (1985). Randomized algorithms. [O'hEigeartaigh *et al.*, 1985], 89-105.

M. O'hEigeartaigh, J.K. Lenstra, A.H.G. Rinnooy Kan (eds.) (1985). *Combinatorial Optimization: Annotated Bibliographies*, Wiley, Chichester.

C.H. Papadimitriou (1985). Computational complexity. [O'hEigeartaigh *et al.*, 1985], 39-51.

ACADEMIC - INDUSTRIAL COLLABORATION IN NUMERICAL ANALYSIS

S. McKee
Department of Mathematics
University of Strathclyde
26 Richmond Street
Glasgow G1 1XH

ABSTRACT. This paper will deal with the author's experience of academic-industrial collaboration. In particular it will discuss the University Consortium for Industrial Numerical Analysis (UCINA), which is a consortium of numerical analysts from a group of British universities, coordinated by the author over a period of six and a half years until 1 February 1986. His experience largely evolves from his involvement with this consortium. Thus a description will be given as to how UCINA was set up and to its general framework, its administration and funding. Details of typical problems tackled by UCINA will also be described.

1. THE INDUSTRIAL PROBLEM

The application of mathematical or numerical techniques to an industrial problem is not straightforward. The industrialist will not be a mathematician and will not be aware of what mathematics can and cannot do. There must be basic qualitative understanding of the physics of the problem. If this does not exist, then the mathematician has little further to contribute for he needs the basic physics or chemistry to build his mathematical model. It is this model which transforms the qualitative information into firmer quantitative information which, in turn, enables the industrialist to make a more positive decision.

The industrialist will often have a problem which he may well state in rather vague terms. How deep is a crack in a pressure water reactor? In fluid-structure interaction what should the spacing be between rigid struts attached to a metal plate to ensure that no longitudinal waves resulting from a point excitation travel along the plate? In rotor stacking of compressor and turbine blades in an aero-engine can they be rotated so as to maintain the overall tolerance below a given acceptable level?

It is first necessary to discuss the matter closely with the scientist in order to home in on a particular, well defined, physical problem or phenomenon or some aspect of it. This having been done, a mathematical model, usually a continuum model, has to be built. Standard mathematical techniques such as non-dimensional analysis (which often enables small terms to be neglected), perturbation analysis and transform techniques can be employed to reduce the size of the problem, provide further understanding and occasionally yield a solution sufficient for the industrialist's needs. Nonetheless in the majority of cases, a numerical solution will be required and so it is necessary to devise some numerical model, or equivalently, to reduce the continuum model to some discrete model, often through finite difference or finite element methods. Even after often substantial software development the problem is then not solved; the results must be interpreted and, if these are extensive, special post-processing may be needed. Their interpretation may not agree with either physical intuition or the results of experiments in which case a control loop requires the mathematician to return to the various stages described above. Assuming that there are no programming errors, that the discrete model has been implemented with sufficient precision and that the mathematics has been carried out correctly, one has to ask if the mathematical model fully describes those aspects of the problem under consideration. If not, a new mathematical model must be designed and the process restarted.

2. UNIVERSITY CONSORTIUM FOR INDUSTRIAL NUMERICAL ANALYSIS

The University Consortium for Industrial Numerical Analysis, known by the acronym of UCINA, is an umbrella organisation consisting of numerical analysts from six universities in the south of England and Wales and those from the Division of Information Technology and Computing at the National Physical Laboratory at Teddington. The universities involved are Bath, Brunel, Oxford, Reading, Swansea and Imperial College. The aims of UCINA are not only to bring to the attention of university mathematicians problems of current importance to industry but also to disseminate recent advances and new techniques to industrialists.

UCINA began 1 October 1979 and was funded by the Science and Engineering Research Council until 30 September 1984. It is now financed directly by industry with important support from the Department of Trade and Industry. The funding covers the salary costs of the coordinator. The organisation of UCINA is as given in diagram 1.

There was close liaison between the author and Dr John Ockendon of the Oxford Study Groups with Industry, built upon a good personal relationship and a free and helpful exchange of information and projects. UCINA's association with the Numerical Algorithms Group (NAG) is as yet at the informal stage. The presence of these two groups was, however, instrumental in the UCINA coordinator being based at Oxford.

The steering committee consists of senior representatives from the six universities, the coordinator and Mr Miller of NPL. The committee sets the policy and the direction of UCINA; the coordinator is responsible for its operation.

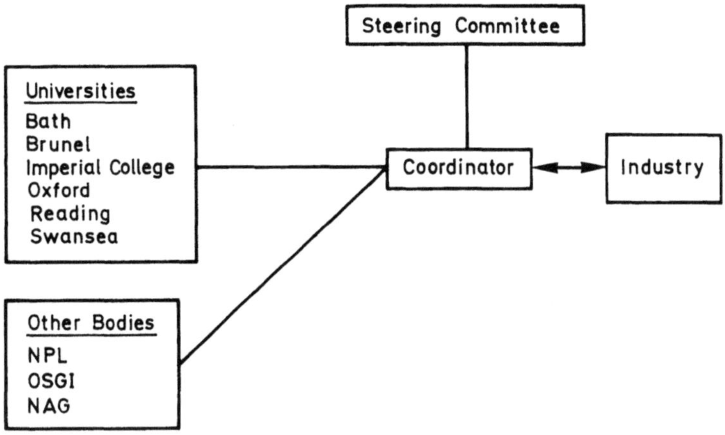

diagram 1

(NPL = National Physical Laboratory;
OSGI = Oxford Study Groups with Industry;
NAG = Numerical Algorithms Group).

3. HISTORICAL BACKGROUND

In 1977 a report of the Numerical Analysis Panel, under the chairmanship of Professor K W Morton, (see Morton (1977)) recommended that proposals should be invited for regional groupings of universities whose objectives would be to improve communication between university numerical analysts and industry and to increase the effectiveness of numerical analysis in solving practical problems. Encouraged by Professor K W Morton, Professor L Fox submitted an application to the SERC for funding for three years to support a project coordinator, a part-time secretary and secretarial and travel expenses. The five original universities were chosen because of their geographical proximity and because a link already existed in the form of joint research student symposia which took place, and still do, twice a year at one of the five universities. Thus on 1 October 1979 the author was appointed as the coordinator. After three years the SERC extended the funding for a further two years on the condition that UCINA become self-funding thereafter. How this was done and the form it took is described briefly in Section 10.

However the background goes back even further than the 1977 Numerical Analysis Panel report. In November 1965, the Council of the Royal Society decided to set up an ad hoc committee to examine post-graduate training in science and technology in the United Kingdom. The report of the mathematics sub-committee which appeared in 1968 (see Blackett (1968)), stressed the necessity for the orientation of research in applied mathematics towards relevant real problems; it also argued that graduate students and faculty members should have experience in formulating problems. It was partly in response to this report that in 1968 Professor L Fox and Dr A Tayler (Fox and Tayler (1969)) set up an experiment in academic-industrial co-operation, which later became known as the Oxford Study Groups with Industry (OSGI). Problems were solicited from industrial mathematicians, engineers and scientists and during one week at the end of Hilary Term faculty members, graduate students and industrialists worked together at Oxford formulating the problems and attempting to solve them. As the mathematics associated with the problems evolved, lectures on relevant topics such as dimensional analysis and integral transforms were given. Of course analytic solutions, in the form of an asymptotic expansion or appropriate solutions obtained through perturbation analysis, were seldom adequate and consequently numerical techniques were needed. These meetings have been held every year since and numerical analysts became more involved. It was therefore natural, when UCINA came into being, that the project coordinator should be based at Oxford.

4. CONTACT WITH INDUSTRY

The first task of the coordinator was to build up contacts with industry. Dr J Ockendon provided an initial mailing list which has now been extensively increased. This was done in part through Computer Survey which details all United Kingdom firms with computers and the uses to which they put them. Conference mailing lists and lists of new members of the Institute of Mathematics and its Applications were also helpful. However old students and word-of-mouth contacts have invariably proved to be the most useful.

A large number of visits have been made to firms and Government Establishments. The purpose of these visits is to solicit problems for the two or three open meetings held generally at Oxford or Imperial College. At these meetings the industrialists present their problems while the academics offer their solutions to earlier problems. Initial formulation, so that a problem is suitable for subsequent numerical analysis, is usually done by the coordinator.

The industrialist is not encouraged to believe that his problem can be solved immediately. Rather a contact is made and the coordinator arranges follow-up meetings with interested colleagues. On occasions firms or Government Establishments do not like to give a public presentation. In this case the coordinator selects the appropriate academic colleagues and a private meeting is arranged.

5. INDUSTRIALLY ORIENTATED M.Sc. COURSES

The universities of Brunel, Oxford and Reading offer M.Sc. courses in applied numerical analysis. Brunel's emphasis is on continuum mechanics while Reading tends to specialize in computational fluid dynamics. The early experience with OSGI confirmed original expectations that the solution of problems in the applied sciences and industry requires a combination, in various degrees, of skills in problem formulation, analytical treatment and numerical work and this led in 1979 to the formulation of a new M.Sc. course in mathematical modelling and numerical analysis. This vocationally orientated M.Sc. requires all students to sit two core papers, one in mathematical modelling and one in numerical analysis, in addition to two further optional papers in either one area or the other. A dissertation is also written often on some real industrial problem arising from OSGI or UCINA's contacts. These theses frequently provide an initial preliminary study prior to tendering for a major contract.

6. UCINA'S EDUCATIONAL ROLE

UCINA's role is primarily educational in a broad sense. A large proportion of the M.Sc. and Ph.D. students from the five universities write dissertations on some application of numerical analysis to industrial problems. The problems that the coordinator obtains from industry are therefore fed back into the educational programme. The essence of some of these dissertations are given in M.Sc. lectures in subsequent years and occasionally find their way into the undergraduate course. The graduates thus produced are therefore well prepared for a life in industry and do not have to be "re-educated".

Industrial problems also provide faculty staff with interesting problems and new lines of research. If the academic is unable to tackle it himself he may, perhaps with some help from the coordinator, apply for a CASE studentship* or for funding for a research assistant through one of the many Science and Engineering Research Council (SERC) schemes. Occasionally industry is happy to meet the total cost of a project. In this way faculty staff are kept in touch with real problems. On the other hand because of the number of numerical analysts involved UCINA is able to offer advice on a wide range of topics. Informal consultative meetings are held frequently and it has been found that academic staff are happy to listen to the industrialist and offer him their advice freely.

Computational modelling is increasing at the expense of physical modelling and, as industry attempts to solve larger and more difficult problems, their need for efficient and robust algorithms grows. It is here that numerical analysis can contribute most and UCINA has the opportunity to promulgate new techniques such as sparse matrix methods, conjugate gradients and multigrid. Instructional courses, specially tailored to meet a firm's particular requirements are given to industry. This usually involves a team of two or three academics. They visit the firm, having read some of their internal reports, give a one hour talk or two 40 minute talks on their speciality, answer questions and

*A CASE studentship is one jointly funded by the Science and Engineering Research Council and a particular industry or Government Establishment.

during long tea, coffee and lunch breaks talk informally with the industrial participants.

7. AREAS OF RESEARCH

UCINA is not only involved in trouble-shooting but also in co-ordinated research programmes. One-off problems can be interesting but resources need to be replenished and so co-ordinated research programmes are necessary. Thus at Brunel the major effort is in continuum mechanics and in particular in non-linear elasticity whereas at Reading the expertise is in computational fluid dynamics and optimal control. The recent formation of the Institute of Computational Fluid Dynamics (ICFD), now shared between Oxford and Reading, has added to Oxford's research capabilities which before were scattered across most of numerical analysis with perhaps special emphasis on the numerical solution of free and moving boundary value problems.

UCINA has received a wide variety of problems. The Central Electricity Generating Board (C.E.G.B.) has provided problems on fluid flow, galvanic corrosion, non-destructive testing, ageing in steel, particles flowing in a turbulent fluid, optimal control and vibration in a gear box. Rolls Royce presented problems in turbulent fluid flow, numerical coordinate generation and eigensolution techniques. Other firms have mathematical problems in a specific area: for instance, Pilkington Brothers are almost exclusively concerned with the flow of molten glass while the Admiralty Research Establishment (A.R.E.) are most interested in flow past smooth bodies.

Studies associated with energy have played an important part. For example, the United Kingdom Atomic Energy Authority placed a contract through UCINA at Oxford University to look after, run and interpret the results produced by an American code, modelling the behaviour of nuclear fuel elements under various coding regimes. Meanwhile at Reading University Dr Nancy Nichols was, and still is, studying control strategies for power generation from tidal flow associated with the proposed Severn barrage. This work is continuing for a further three years funded by a joint SERC/CEGB grant*. Natural gas from the North Sea will eventually become exhausted and there is, consequently, an interest in considering the possibility of building large plants to extract gas from coal. The coal is burnt at the bottom of a tall cylindrical drum and a study of the slow granular flow of the coal is a necessary part of understanding the process with a view to predicting what will happen when the small pilot plant is increased ten fold. A preliminary study of numerical methods for solving Spencer's granular flow equations was undertaken.

UCINA is also involved with the new technologies. Numerous problems in the field of semiconductors are being considered. In particular, UCINA held a joint contract with the General Electric Company (G.E.C.), Plessey and the Standard Telecommunication Laboratories (S.T.L.) to study process modelling of semiconductor devices. This is

*CEGB = Central Electricity Generating Board

now funded directly by Alvey, a specially promoted government scheme to encourage research in information technology. Mathematically this requires the solution of coupled highly nonlinear diffusion equations on a difficult geometry with two moving boundaries. The intention is to put together a sophisticated computer program which will help engineers to design smaller silicon chips. This work is being done mainly at Reading under the supervision of Dr Michael Baines with a little input from the coordinator. Another aspect of the new technologies is the manufacture of optical fibres. Mathematically this requires the solution of the low Reynolds number Navier-Stokes equations with a time dependent free surface. This has been successfully accomplished by Dr Joyce Aitchison and one of her students at Oxford and has led to a better assessment of the process. Oxford Instruments manufacture superconducting magnets which are then used in scanners in hospitals. Screening of these magnets is important since a pacemaker even two rooms away or in the floor above can be thrown out of order. The major difficulty here arises from the requirement that the metal plate, which is the screen, must necessarily be thin which, in itself, causes numerical difficulties. This work was undertaken at Imperial College by Dr Michael Bernal and his student.

UCINA has an interest and expertise in the broad area of elasticity and plasticity. A substantial contract with the National Gas Turbine Establishment (now part of the Royal Aircraft Establishment) to study real time simulation of the stress on turbine blades was completed at Brunel University under the direction of Professor John Whiteman of BICOM. A problem from BICC concerned the vertical unreeling of a wire from a coil on the ground as fast as possible without snagging. Although this can be modelled as a two point boundary value problem, the boundary conditions are not clear and all the standard NAG routines break down. This was the starting point that led to considerable analysis and the development of further models. This problem is not yet fully resolved. Stress in timber is another important area and UCINA acts as a consultant to the Timber Research and Development Association. Finally UCINA holds a contract with the National Coal Board to study stress and displacement in soft rocks.

Optimization is also a field in which UCINA is involved. The problem of optimal tool packing was tackled. This concerned the best way to pack a given consignment into boxes of a given size subject to certain constraints. While simple integer programming models were derived, the real problem could only be solved in a number of steps which grew exponentially with the number of tools. As is well known in complexity theory this is called np-complete. The negative conclusion that it is not feasible to solve the full problem gave some reassurance to the company concerned that their existing methods of packing were reasonable.

These are only a small selection of the many problems that have been considered by UCINA during the six years of its operation. Some of these problems and others have appeared in detail in a book entitled "Industrial Numerical Analysis", edited by the coordinator and Dr Charles Elliott, and published by Oxford University Press in December 1985.

8. A SELECTION OF INDUSTRIAL PROBLEMS

This section will deal superficially with four typical industrial problems. The manufacture of glass rods and fibres gives rise to an elliptic problem with a free surface while the manufacture of semi-conductors requires the solution of a nonlinear diffusion equation. The migration of particles in a turbulent fluid involves the solution of an integro-differential equation and, in order to understand granular flow, a system of nonlinear hyperbolic equations needs to be solved.

8.1 Glass fibre updraw process

8.1.1 Introduction.

Pilkington Brothers were interested in a more precise quantitative assessment of this process of manufacturing glass rods and fibres. The process is as shown in diagram 2.

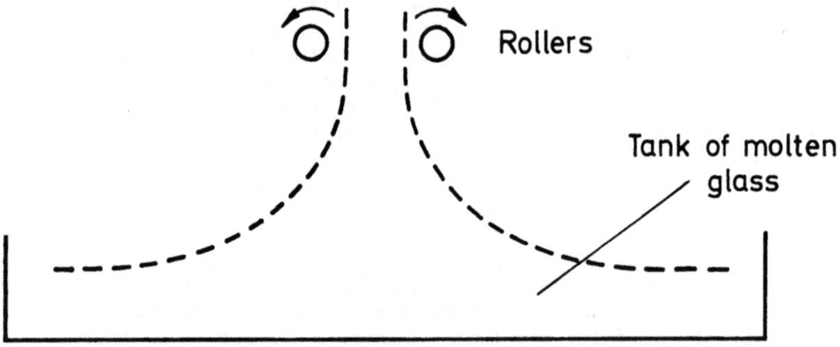

diagram 2

The two crucial assumptions are that the inertia terms can be ignored since Re is small and that the problem can be treated as isothermal. The latter assumption might appear doubtful. However a comparison between experiment and the numerical result of the computational model showed this to be justified. This problem was presented at the first UCINA meeting held on 17 January 1980 and was tackled successfully by Dr Joyce Aitchison and her student, Dr Colin Saxelby. (See Saxelby and Aitchison (1985), Saxelby (1981)).

8.1.2. Mathematical model. The governing differential equations are the Navier Stokes equations without the inertia terms:

$$\frac{\partial}{\partial x}(t_{xx}) + \frac{\partial}{\partial y}(t_{xy}) = 0$$

$$\frac{\partial}{\partial x}(t_{yx}) + \frac{\partial}{\partial y}(t_{yy}) - \rho g = 0$$

together with the continuity equation

$$\frac{\partial u}{\partial x} + \frac{\partial v}{\partial y} = 0$$

where u,v are respectively the horizontal and vertical components of velocity, ρ is the density (assumed constant), and g is the acceleration due to gravity. The stresses t_{xx}, t_{xy}, t_{yx}, t_{yy} are defined by

$$t_{xx} = -p + 2\mu \frac{\partial u}{\partial x}$$

$$t_{xy} = t_{yx} = \mu(\frac{\partial u}{\partial y} + \frac{\partial v}{\partial x})$$

$$t_{yy} = -p + 2\mu \frac{\partial v}{\partial y}$$

where p is the pressure and μ the viscosity which is taken to be constant as a consequence of the isothermal condition. Note that a steady state solution is being sought.

8.1.3 Boundary conditions. It is convenient to define the boundary tractions

$$T_x = n_x t_{xx} + n_y t_{xy}$$

$$T_y = n_x t_{xy} + n_y t_{yy}$$

where (n_x, n_y) is the outward normal to the surface.

The boundary conditions are then given in diagram 3.

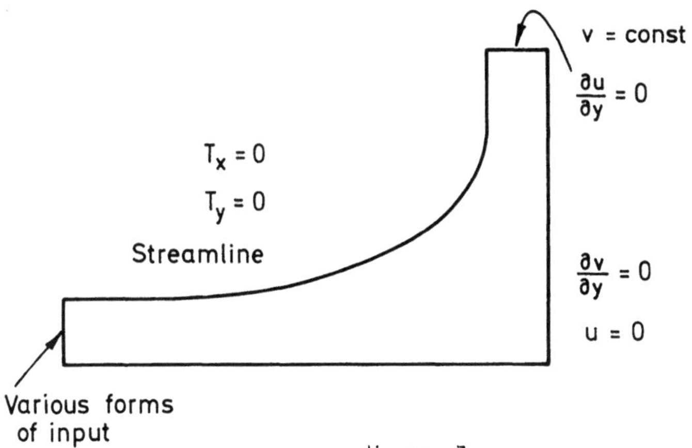

diagram 3

Note the three boundary conditions on the free surface: this is necessary for the well-posedness of the problem since the free surface is itself unknown and must be determined.

8.1.4 The method of solution. The Galerkin finite element method using an adaptive triangular grid was employed. The fluid velocities u,v were approximated by piece-wise quadratic functions while the pressure p was approximated by piece-wise linear functions. The algorithm can be summarized as follows:

1. Guess the position of the free surface.
2. Solve in the fixed region (using the stress conditions on the free surface).
3. This will not satisfy the streamline condition in general. Use this condition to determine a new position of the free surface.
4. Go to 2 until the error between two successive free surfaces is sufficiently small.

8.2 Process modelling of semiconductors

8.2.1 Introduction.

In the manufacture of semiconductors an initial dopant, which might for instance be arsenic or boron, is implanted in the silicon chip. Annealing then takes place, diffusing the implant nonlinearly. At the same time a silicon dioxide layer grows. Depending on the temperature of the anneal this can be treated as an elastic or viscoelastic material. A nitride cap (Si_3N_4) covers the silicon dioxide. Since the cap has some degree of flexural rigidity and this has the effect of deforming the silicon dioxide there is a moving boundary between the SiO_2 and Si, another between the SiO_2 and Si_3N_4 and a free surface on the SiO_2. There is interest in the implantation of up to 4 dopants. Account must also be taken of clusters of implantations; these clusters are however not mobile and so have an effect on the diffusion.

UCINA, through the University of Reading, holds a contract to produce a computer program which will simulate the diffusion of the dopants, the growth of the SiO_2 region and clustering. This is being supervised by Dr Michael Baines. In this code the SiO_2 is assumed non-elastic, that is, rigid.

The formation and the change in the shape of the viscous SiO_2 region is the subject of a mathematical study by Dr Alan Tayler and his student, John King, who is funded by a CASE award with Plessey. In this section we shall briefly discuss both projects.

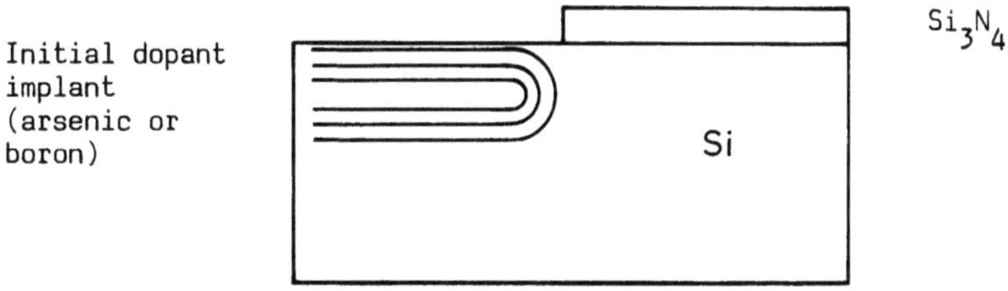

diagram 4

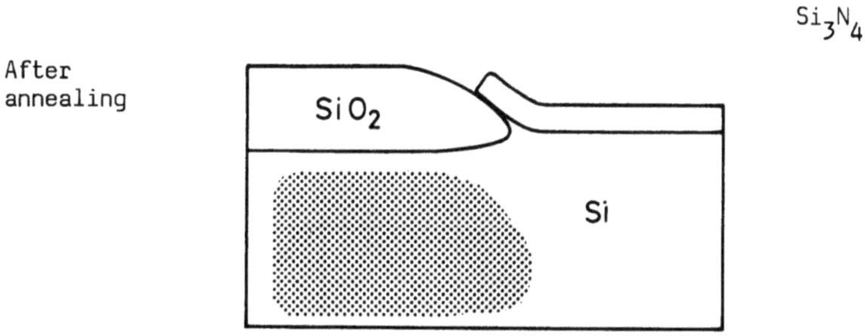

diagram 5

8.2.2 Mathematical problem of the dopant diffusion. The simplest problem is given in diagram 6.

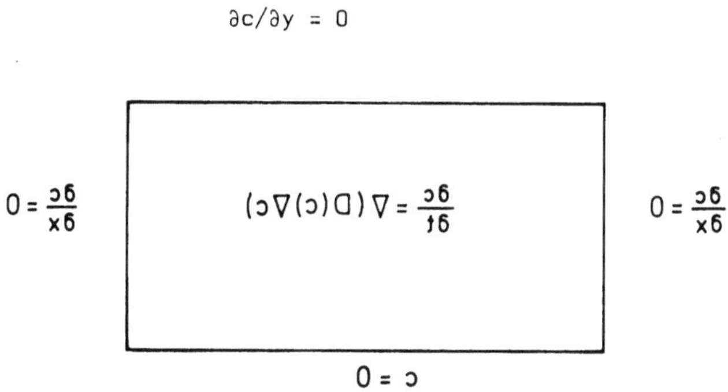

diagram 6

The concentration of the dopant is denoted by c while the nonlinear diffusion coefficient is given by

$$D(c) = D_0\left(\frac{1+\beta c/n_i}{1+\beta}\right)\frac{(1+c)}{(c^2+4n_i^2)^{\frac{1}{2}}}$$

where n_i are known (or measurable) constants and D_0 is the known intrinsic diffusion coefficient at a given temperature. Clustering is simply the combining of ions of the dopant with each other and so can be expressed as an ordinary differential equation with a known rate constant.

8.2.3 Numerical technique. A finite element method, using triangular elements in space and Crank Nicolson in time, has been in existence for some time now. At each time step a system of nonlinear equations is required to be solved and this is done very rapidly by means of outer and inner iterations, the latter using a preconditioned conjugate gradient technique.

The efficiency of the code is currently being improved by the application of the ideas of the method of moving finite elements (see Wathen and Baines (1983)). The next stage is to incorporate the moving boundary and the free surface and then the possibility of four diffusing dopants.

Existing codes which are essentially 1-D are SUPRA, SUPREM, ICECREM. A good acronym will be required for this code when it is complete.

8.2.4 Mathematical model of silicon isolation oxidation. The non-dimensionalized equations for the slow viscous flow of the silicon dioxide are as given in diagram 7.

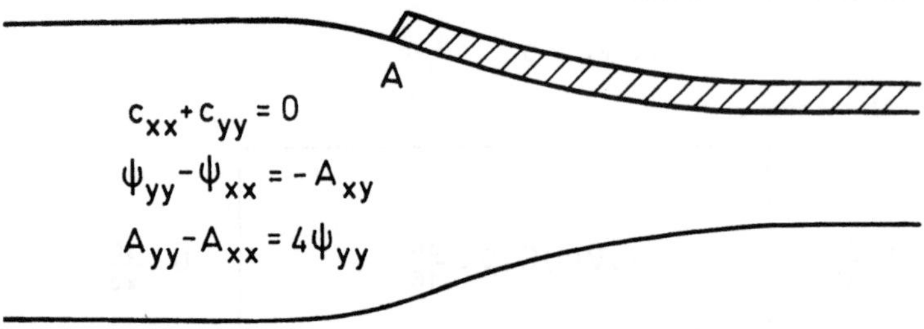

diagram 7

Note that here c denotes the concentration of oxygen and that $\frac{\partial c}{\partial t} \sim 0$ since the diffusion of oxygen occurs much more rapidly than any change in the moving boundary or free surface. The other dependent variables are the stream function and the Airy stress function. The details of the boundary conditions can be found in King (see also McKee (1984)).

8.2.5 Solution techniques. John King has attacked this problem from many angles. He has, after linearization, obtained a small time solution; he has also obtained the nature of the singularity at A; and an expansion for a small value of one of the parameters has allowed him to get both small and large time solutions. Finally, and perhaps of more interest, he has been able to rewrite the full problem in a boundary integral formulation and a numerical solution is being sought.

8.3 Diffusion coefficient of particles in an isotropic turbulent fluid

One of the interesting aspects of this problem is that it illustrates how numerical analysis can identify an error in classical applied mathematics.

Dr Michael Reeks of Berkeley Nuclear Research Laboratories was interested in the movement of particulate matter in a turbulent fluid. In particular he wished to obtain the asymptotic particle diffusion coefficient (as $t\to\infty$) in an unbounded stationary homogeneous turbulent flow when the particle is subjected to both added mass and Basset history forces as well as the normal Stokes drag. Such information would then allow him to determine the extent of migration of radiative particles within a nuclear reactor by solving the transport equation in the appropriate geometry.

One way of attacking this problem is to attempt to solve a non-linear extension, due to Odar and Hamilton (1964), of the Basset (1988), Boussinesq (1903) and Oseen (1927) BBO equation many times and compute the diffusion coefficient through ensemble averages. Historically this had already been done analytically by Tchen (1947) for the slow motion of a particle in a quiescent fluid. Specifically Tchen's equation was

$$\frac{d\underline{v}}{dt} - \beta\frac{d\underline{u}}{dt} + \alpha(\underline{v}(t)-\underline{u}(t)) + \gamma\int_0^t (t-s)^{-\frac{1}{2}}\left(\frac{d\underline{v}(s)}{ds} - \frac{d\underline{u}(s)}{ds}\right)ds = 0 \qquad (1)$$

where α, β and γ are interrelated constants of motion depending upon the particle and fluid properties. For the case of a spherical particle of diameter d and density ρ_p in a fluid of density ρ_f and dynamic viscosity μ,

$$\alpha = \frac{18\mu}{d^2(\rho_p+\frac{1}{2}\rho_f)}, \quad \beta = \frac{3\rho_f}{2\rho_p+\rho_f}, \quad \gamma = \frac{9(\rho_f\mu/\pi)^{\frac{1}{2}}}{d(\rho_p+\frac{1}{2}\rho_f)} .$$

Odar and Hamilton's nonlinear equation had been derived empirically for larger Reynolds numbers for a sphere moving in an isotropic turbulent fluid. The velocity u(t) of the isotropic turbulent fluid was modelled, following Kraichnan (1970), by fifty terms of its Fourier series with the amplitudes, frequencies and wave numbers all normally distributed.

After designing and analysing a product integration method (see McKee (1979)) to solve the nonlinear version of (1) the results of the computer program were found to differ substantially from the known analytic results, due to Tchen (1947), for the corresponding linear problem. This led us to numerically check the linear results and then when they were also found to differ, to perform our own asymptotic analysis (McKee, Reeks and Stokes (1984)). The results of this analysis, together with the supporting numerical evidence, were then interpreted physically (Reeks and McKee (1984)).

In short these state that the value, as defined by Tchen, for the asymptotic particle diffusion coefficient in an unbounded stationary homogeneous turbulent flow at low particle Reynolds number is in error by an additional term that depends wholly on the existence of the Basset history term. It is usually claimed (Hinze (1959), p.355, Soo (1967), p.33, Boothroyd (1971), p.22 or the original source Tchen (1947)) that the Basset term is unimportant if the density of the fluid is comparable to, or less than, that of the particle. Yet for a sphere of diameter d this extra term is given by

$$(d^2/2\pi\nu_f)\sigma$$

where ν_f is the kinematic viscosity and σ the co-variance of the incipient relative velocity between the particle and the local fluid. Noticeably this extra term does not depend upon the particle density.

8.4 Granular flow

With the life of North Sea gas limited there is a growing interest in gasification, that is, the manufacture of gas from coal burning. This process uses large cylinders and it is important to understand how the coal moves down the cylinder to the combustion area. Thus a numerical simulation of the equations governing the granular flow would be useful. Unfortunately this field is not yet fully understood and there are competing mathematical models. The object of this study is to effect a comparison between Spencer's model (1964) and Josselin de Jong's model (1971) and to attempt to discover which, if either, is the more appropriate for this particular problem. We shall here only discuss Spencer's model.

8.4.1 Mathematical model (Double Shearing model)

If σ_I and σ_{II} denote respectively the major and minor principal stresses and $q = \frac{1}{2}(\sigma_I - \sigma_{II})$ and ψ is the angle between the x-axis and the major principal stresses then Spencer (1964) has shown that q, ψ and the velocity components u and v satisfy the hyperbolic system

$$A\underline{u}_x + B\underline{u}_y = \underline{c}$$

where

$$A = \begin{pmatrix} \cos 2\psi - \operatorname{cosec}\psi & -2q\sin 2\psi & 0 & 0 \\ -\sin 2\psi & 2q\cos 2\psi & 0 & 0 \\ 0 & 0 & 1 & 0 \\ 0 & 0 & -\sin 2\psi & \cos 2\psi - \sin\psi \end{pmatrix}$$

$$B = \begin{pmatrix} \sin 2\psi & 2q\cos 2\psi & 0 & 0 \\ -(\cos 2\psi + \operatorname{cosec}\psi) & 2q\sin 2\psi & 0 & 0 \\ 0 & 0 & 0 & 1 \\ 0 & 0 & \cos 2\psi + \sin\psi & \sin 2\psi \end{pmatrix}$$

and $\underline{c} = (0,\rho g,0,0)^T$,

while ρ denotes the density and g is the usual acceleration due to gravity. This model represents plane stress and assumes that no dilation takes place. A similar set of equations can be written down for cylindrical geometry.

8.4.2 Numerical solution. One of the attractive features of this model is that the equations decouple and it can be shown that the velocity components share the same characteristics as q and ψ. These are illustrated in diagram 8.

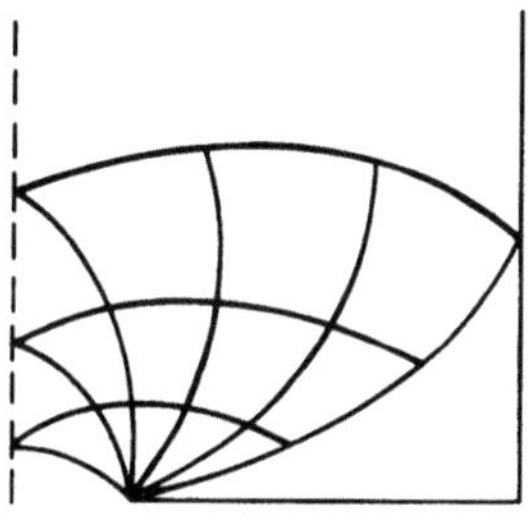

diagram 8

This has now been solved by M Meneguette at Oxford using a simple method of characteristics, although a major headache in this problem is knowing what the boundary conditions should be. The De Josselin de Jong model is more difficult to solve since the characteristics of the stress and velocity field do not coincide. After obtaining the stress field the nonlinear nature of the characteristic relations for u and v require accurate two dimensional interpolation of unevenly distributed points.

9. UCINA TODAY

UCINA is now funded directly by industry through subscription. There are two categories of industrial membership, the first category receiving more rights and privileges than the second. At present there are some fifteen industrial members.

UCINA has a brochure describing the organisation, an extensive set of lecture programmes detailing the courses available and a report series aimed at bringing the latest techniques to the attention of the industrial community.

Category I members receive free copies of the UCINA reports, several one day visits by the coordinator, open invitations to talk about problems at a UCINA meeting, close liaison on any problem presented including several informal follow-up meetings. A one-day lecture programme or a one-day visit by a UCINA consulting team is also offered. Category II members receive a subset of these rights and privileges.

Further earnings are obtained through the specialist lecture service and through a 10% levy on contracts. UCINA also benefits from the Department of Trade and Industry acting as a guarantor for up to 30% of our costs over the first three self-funding years.

10. CONCLUDING REMARKS

A question that is often asked is: "Should universities be involved in this endeavour?" It is the author's belief that we should be, provided it does not take over from our primary role as educational institutions. It is beneficial, not only in allowing us to increase our faculty staff, but, more importantly, in allowing us to assess the relative importance of different areas of our own discipline. For example, out of one hundred or so problems tackled by UCINA members over the last six years, not one dealt with the numerical solution of an initial-value ordinary differential equation and yet a large proportion of the numerical analyst community specialise in this area.

There are, however, dangers. To spend one's time bidding for contracts and solving industrial problems could be counter-productive in the longer term. It is essential that such research is backed up by fundamental research; industry cannot be expected to finance this and funding must come from the appropriate governmental body or research council. It is important, also, not to let Mammon take over; good (applied) mathematics and money are not good bed-fellows. The solution of industrial problems is best done through free and open discussion often using the expertise and skills of as many as a dozen people. As coordinator of UCINA the author has striven to maintain a balance between good numerical analysis and financial viability.

Finally it cannot be stressed too strongly that the academic must be prepared to seek out industry's problems. Indeed academic-industrial collaboration demands certain attitudes from both the academic partner and the industrial partner: to repeat, the academic must be prepared to tackle a wide range of mathematical problems and not just attempt to seek out those problems within his own, often narrow, specialization; the industrialist, on the other hand, must have some sympathy and understanding for the aims of the academic and not require a weak answer, which might be substantially inaccurate or even irrelevant, in the shortest possible time. If this symbiosis can be achieved, it will be mutually advantageous.

REFERENCES

Basset, A.B. A Treatise on Hydrodynamics (Vol.2, ch 5), Deighton Bell and Co., Cambridge (1888).
Blackett, P.M.S. 'Postgraduate education in applied mathematics', Bull. I.M.A. 4, No.14 - 20 (1968).
Boothroyd, R.G. Flowing Gas-Solids Suspensions, Chapman and Hall (1971).
Boussinesq, J. Theorie Analytique de la Chaleur, Vol.2, p.224, Gauthier-Villars, Paris (1903).
Elliott, C.M. and McKee, S. Industrial Numerical Analysis, Oxford University Press, (1985).
Fox, L. and Tayler, A.B. 'An experiment in academic-industrial co-operation': Oxford Study Groups on differential equations, Bull. I.M.A., 5, No.4, 82-84 (1969).
Hinze, J.O. Turbulence (2nd edition), McGraw-Hill Inc., New York (1975).
De Josselin de Jong, G. The double sliding, free rotating model for granular assemblies. Geotechnique 21, 155-163 (1971).
King, J. The Oxidation of Silicon in the Process Modelling of Semi-conductors. D.Phil. Thesis, University of Oxford (to be submitted).
Kraichnan, R.H. Diffusion by a random velocity field. Phys. Fluids 13, No.1, 22-31 (1970).
McKee, S. The analysis of a variable step, variable coefficient linear multistep method for solving a singular integro-differential equation arising from the diffusion of discrete particles in a turbulent fluid. J Inst. Maths Applics. 23, 373-388 (1979).
McKee, S. 'UCINA - a vehicle for academic-industrial collaboration'. Proceedings of the Tenth South African symposium on numerical mathematics, Balito, 1 - 27 July (1984).
McKee, S., Reeks, M.W. and Stokes, A. The asymptotic analysis of particle dispersion caused by random history forces. ZAMM, 64, 560-564 (1984).
Morton, K.W. (Chairman) Report of the Numerical Analysis Panel. Science and Engineering Research Council (1977).
Oseen, C.W. Hydrodynamik, Leipzig (1927).
Reeks, M.W. and McKee, S. The dispersive effects of the Basset history forces on particle motion in a turbulent flow. Phys. Fluids, 27 1573-1581 (1984).

Saxelby, C. and Aitchison, J. A numerical model of the glass sheet and fibre updraw processes. Industrial Numerical Analysis. (Eds. C.M. Elliott and S. McKee) (1985).

Soo, S.L. Fluid Dynamics of Multiphase System, Blaisdell (1967)

Spencer, A.J.M. A theory of the kinematics of ideal soils under plain strain conditions. J. Mech. Phys. Solids 12, 337-351 (1964).

Tchen, T.M. Mean Value and Correlation Problems Connected with the Motion of Small Particles Suspended in a Turbulent Fluid, Ph.D. Thesis, University of Delft (1947).

Wathen, A.J. and Baines, M.J. On the structure of the moving finite element equations. Numerical Analysis Report 5/83. Department of Mathematics, University of Reading (1983).

THE OPTIMAL FORM OF DIAMOND HEAT SINKS

J. Molenaar, G.W.M. Staarink
Wiskundige Dienstverlening
Faculty of Science, University of Nijmegen
Toernooiveld, 6525 ED Nijmegen
The Netherlands

ABSTRACT. The heat diffusion properties of heat sinks used in industry to remove the thermal energy in small electronic devices, is analysed. The systems studied consist of diamond and copper. The thermal conductivity of diamond is temperature dependent whereas in copper this quantity is constant in the relevant temperature range. This obstructs an analytical approach. It is shown that the problem can be reduced to two separate Laplace problems in the diamond and copper regions respectively. The application of standard techniques such as finite differences and Green function methods are dealt with in some detail. Special attention is paid to the non standard problem of matching the solutions in the separate regions at their interface. In the results suggestions are given how the geometry of the diamond can be chosen such that the system reduces the maximum temperature for a given produced amount of heat as much as possible.

1. INTRODUCTION.

In various microscale techniques a considerable amount of thermal energy is continuously produced within a small volume. This energy has to be rapidly removed, because otherwise the increasing temperature will cause malfunctioning or even destruction of the device. To that end a special type of diamond is applied with extremely useful thermal properties (see [2] and [3]). Though being an electric insulator its thermal conductivity in the range of 0–200°C is about four times as high as it is in copper. By mounting the device on a commodious diamond the energy is fast spread over a larger region so that the maximum temperature does not increase too much. To save money, copper is used for the further heat transport to the surroundings.

The thermal conductivity of diamond is, contrary to that of copper, temperature dependent and decreases with increasing temperature. Therefore the heat-diffusion equation is in this case nonlinear. Because of this it may happen that the produced amount of heat per unit of time is too large for the geometry under consideration. Then no stationary transport of heat is possible and the temperature will increase infinity. In view of this the **PURPOSE** of the project is :

> **Calculation of the stationary temperature distribution in diamond-copper systems for several geometries as a function of the applied heat flux.**

In practice two essentially different geometries are used: either the diamond is mounted on top of the copper or it is sunken into the copper block and thus replaces a piece of copper. In this study the efficiency of both geometries will be compared. After the introduction we present the modelling of both systems in § 2. For the numerical evaluation an iterative scheme has to be applied, which is dealt with in § 3. In this scheme use is made of the Green function method. The detailed application of this method is described in § 4. The emphasis of this paper is on the presentation of the applied techniques rather than on a detailed survey of the computed results. In § 5 the most striking results are commented. After the references we give in the appendix some details of the 1-dimensional case, which illustrate some interesting points of the problem.

Throughout the paper the following units are used: length in centimeter, power in Watt and temperature in centigrade with respect to 300 K (roomtemperature).

2. MATHEMATICAL PHYSICAL MODEL.

In this section we describe the mathematical physical model to be used.
The form of the diamond in the applications may vary considerally. In the following we shall pay attention to systems with axial or block symmetry. Within this restriction we shall study two kinds of geometries (see fig. 1 and 2):

1) The diamond is sunken in a large copper block (geometry 1, fig 1).

2) The diamond is mounted on top of a large copper block (geometry 2, fig. 2).

For both geometries we make the following assumptions:

a) The copper block is so big compared to the diamond, that it may be considered as being infinitely big; at infinity the block is at room temperature.

b) The heat transfer to the air is so small compared to the heat transport via the diamond and the copper that it may be neglected.

c) There is a continuous and homogeneous heat flux across the device-diamond boundary, denoted by FIO.

d) In the relevant temperature range the thermal capacities and the densities of the materials are constants.

e) Possible heat resistances at the device-diamond and the diamond-copper boundaries, denoted by FIO and DCB respectively, may be neglected.

Since we are only interested in the stationary solution of the problem, in both materials the heat diffusion equation (see [5] and [7]) reduces to:

$$\nabla(k(T)\nabla T(\mathbf{r})) = 0$$

For the thermal conductivity coefficient $k(T)$ we have (see [4] and [5])

$$\text{in diamond: } k(T) = k_1(T+300)^{-\beta}, k_1 = 27530\ Watt.(centigrade)^{(\beta-1)}/cm, \beta = 1.26 \tag{1}$$

$$\text{in copper}: k(T) = k_0 = 3.87\ Watt/centigrade/cm .$$

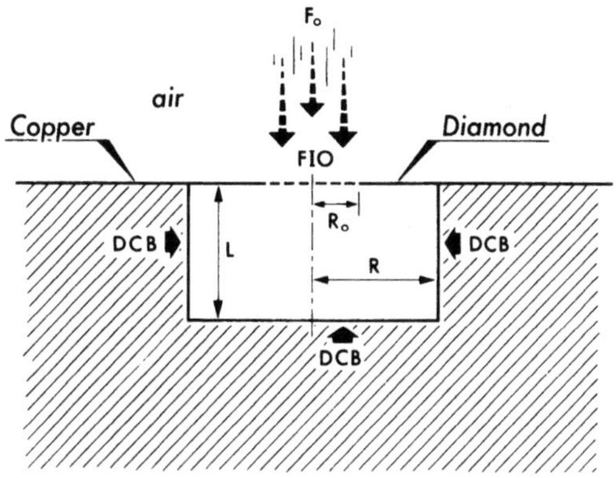

Figure 1: Vertical intersection of geometry 1.
The diamond cylinder, with height L and radius R, is sunken in the very big copper block. From a heat source a continuous flow of heat F_0 enters the diamond through the circular opening FIO, with radius R_0. The diamond-copper interface is denoted by DCB.

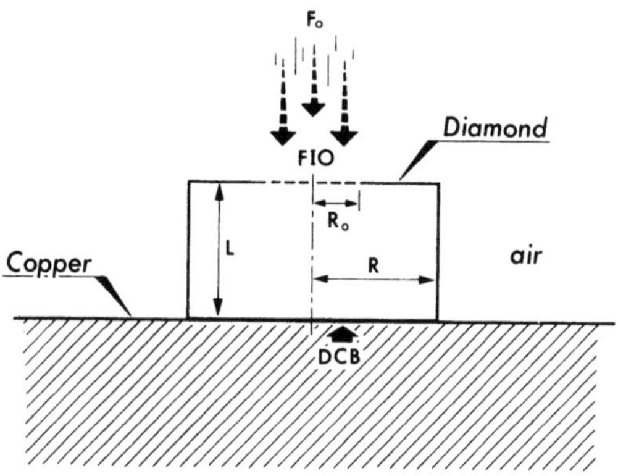

Figure 2: Vertical intersection of geometry 2.
The diamond cylinder is mounted on top of the copper block. The notation is as in fig.1.

Introducing a function $U(T)$ with no actual physical meaning by

$$U(T) = \frac{1}{1-\beta}(T+300)^{1-\beta} \tag{2}$$

the diffusion equations in both diamond and copper reduce to the Laplace equations for U and T respectively:

$$\begin{aligned} &\text{in diamond} : \Delta\, U = 0 \\ &\text{in copper} \quad : \Delta\, T = 0\,. \end{aligned} \tag{3}$$

To complete the mathematical statement of the problem we have to specify the boundary conditions in terms of temperature distributions and heat flux densities. The heat flux in a point on a surface, normal to this surface, is given by:

$$F = -k(T)\frac{\partial T}{\partial n} \tag{4}$$

with the derivative taken in the outward direction normal to the surface.

From the assumptions we have the following conditions:

In **diamond** :

i) Across the *FIO* F is homogeneous and given by

$$F = \frac{F_0}{\pi R_0^2} \;\; or \;\; \frac{F_0}{(2R_0)^2} \tag{5}$$

in case of axial or block symmetry respectively, where F_0 is the total amount of heat per unit time across the *FIO*.

ii) At the diamond-air boundaries it holds that $F = 0$.

In **copper** :

i) At infinity it holds that $T = 0$.

ii) At the copper-air boundary it holds that $F = 0$.

At the diamond-copper boundary **DCB** the conditions are:

i) The T distribution computed in copper and the U distribution computed in diamond should be related by equation (2). In contrast with the flux condition below this relation is nonlinear. It can easily be seen that this obstructs an analytical solution of the problem.

ii) The fluxes at the *DCB* , computed in the diamond and the copper respectively , must balance. Using equations (1), (2) and (4) we may express this requirement as:

$$k_0 \frac{\partial T}{\partial n} = -k_1 \frac{\partial U}{\partial n}.$$

To obtain T, U and F at the *DCB* from these conditions we adopted a numerical iteration scheme, which is described in the next section.

In the 1-dimensional case it is straightforward to solve the model outlined above is analytically solvable. This case is dealt with in the appendix because it is, though not realistic, still illuminating as far as the important features of the problem are concerned.

3. COMPUTATIONAL ASPECTS.

In the numerical computation of the T and F distributions on the *DBC* an iterative procedure is applied, in which the Laplace equations (3) are solved again and again. We shall restrict the presentation of this scheme to geometry 1. Because the used library (see [1]) can only deal with regularly shaped regions, the copper block has been divided into the regions I and II, between which an artificial copper-copper boundary *CCB* is introduced. See fig. 3, wherein the diamond region is denoted by III.

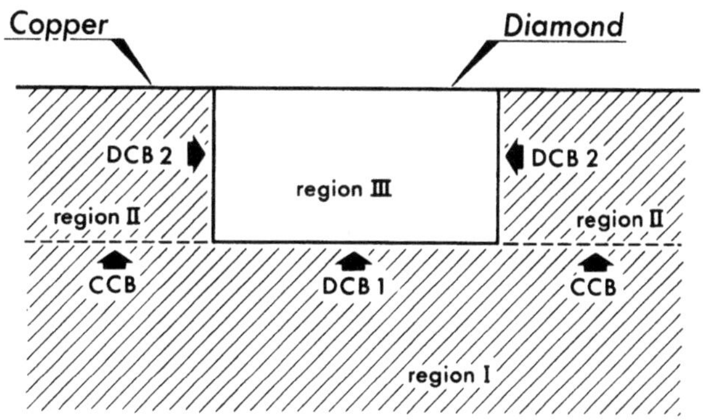

Figure 3 : Division of the copper block of geometry 1 into the regions I and II with artificial copper-copper boundary CCB.

During the ith cycle the approximations T_i and F_i for the T and F distributions on the interfaces $DCB1$, $DCB2$ and CCB are improved. The procedure starts with:

1) Choose an initial flux distribution at the interfaces.

Assume that at the start of cycle i the flux distribution F_i is known. Then this cycle consists of the following steps:

2) Compute in region I the T profile T_i at the $DCB1$ and the CCB.

3) Compute in region II, given the F_i at the $DCB2$ and T_i at the CCB, the T distribution. This directly yields the T_i profile at the $DCB2$. Also compute a new flux $\bar{F}_i$ at the CCB.

4) Compute in region III the U distribution, given the T_i profile at the $DCB1$ (from step 2) and the $DCB2$ (from step 3).

5) Compute from the result of step 4 new fluxes $\bar{F}_i$ at the $DCB1$ and $DCB2$.

6) Compute the new approximation F_{i+1} at the interfaces by the formula

$$F_{i+1} = \gamma F_i + (1-\gamma)\bar{F}_i.$$

7) If F_i and F_{i+1} do not differ more than a given tolerance then stop , else go to step 2.

Remarks on the respective steps.

step 1:
The choice of the initial flux in step 1 is quite arbitrary, provided that the condition of flux neutrality in diamond is satisfied. In physical terms this condition reflects that no heat is generated or lost in the diamond. In practice a homogeneous flux appeared to be a good choice at the $DCB1$ and $DCB2$. At the CCB a vanishing initial flux satisfied.

step 2:
In this step we use the Green function method. The application of this method in our case is explained in the following section.

steps 3 and 4:
To solve the Laplace equations (3) in regions II and III with the given boundary conditions, we used the appropiate subroutines from the FISHPACK library ([1]). These routines compute the solution on a given grid, using a finite difference method. Region II extends to infinity. To avoid numerical problems we applied the transformation $r \rightarrow 1/r$ with r the radius of the cylinder in polar coordinates. By this transformation region II is mapped into a region quite similar to III. The consequently transformed equations with the appropiate boundary conditions are solved using a routine from [1].

steps 3 and 5:
Once the U or T distribution is known at some grid, the flux F across the boundaries is computed by numerical differentiation of U or T in the direction normal to this boundary.

step 6:

In step 6 a weighted mean from F_i and $\tilde{F}_i$ is taken to obtain F_{i+1}. If γ is chosen too small the succesive F_i show an oscillating behaviour and the process does not converge. For each geometry and F_0 value one can find an optimal value for γ for which the convergence rate is greatest. In practice a γ value of 0.95 appeared to satisfy in all cases.

4. GREEN FUNCTION METHOD.

In this section we deal in more detail with the calculations in step 2 of the applied iteration scheme, restricting ourselves to geometry 2.

The size and the boundary conditions of the copper block are such that application of the Green function method is very appropriate. We shall outline that in this method the calculations reduce to integrating the product of the prescibed F distribution and the Green function over the DCB. For a general Green function approach we refer to [9].

For our purposes it suffices to start with the following identity of Green for the Laplace operator Δ within a volume V with surface S.

$$\int_V dV(f\Delta g - g\Delta f) = \int_S dS(fg_n - f_n g)$$

The functions f and g are twice differentiable within V. The derivatives along the outward direction normal to S are denoted by f_n and g_n.
Let S consist of two complementary parts S_1 and S_2. In copper we have to solve the problem:

$$\Delta T = 0\ ,$$

$$T \text{ prescribed at } S_1\ ,$$
$$T_n \text{ prescribed at } S_2\ .$$

The corresponding Green function G is defined as the solution of:

$$\Delta G(\mathbf{r},\mathbf{r}') = \delta(\mathbf{r}-\mathbf{r}')$$

$$G = 0 \ at\ S_1\ ,$$

$$G_n = 0 \ at\ S_2 .$$

The boundary conditions hold for one of the arguments of G but, because G is symmetric in its arguments, we need not a more explicit notation.
With the substitutions $T = f$ and $G = g$ the given identity of Green yields an explicit expression for T in V:

$$T = \int_{S_1} dS\ TG_n - \int_{S_2} dS\ T_n G$$

This expression reduces further if we specify the boundary conditions more in detail. In copper S_1 corresponds with infinity, where T is assumed to vanish. S_2 corresponds with the bounding plane of the copper block on which the diamond is mounted. From assumption b) we have that on this boundary the flux density vanishes outside the DCB. From relation (4) it follows that the same holds for T_n. Using this relation we may express T in terms of F.

$$T(\mathbf{r}) = +\frac{1}{k_0} \int_{DCB} d\mathbf{r}' G(\mathbf{r},\mathbf{r}') F(\mathbf{r}')$$

It remains to find G. This problem is most easily solved by the method of images (see [3], [4], [5]), which yields

$$G(\mathbf{r},\mathbf{r}') = \frac{1}{4\pi}\left\{\frac{1}{|\mathbf{r}-\mathbf{r}'|} + \frac{1}{|\mathbf{r}-\mathbf{r}''|}\right\}$$

where $\mathbf{r}''$ is the mirror point of $\mathbf{r}'$ with respect to the boundary. Because we only need G with $\mathbf{r}'$ at the DCB and thus in this plane we may use G in the form:

$$G(\mathbf{r},\mathbf{r}') = \frac{1}{2\pi} \frac{1}{|\mathbf{r}-\mathbf{r}'|}, \qquad \mathbf{r} \in DCB.$$

We shall apply these equations for both the axial symmetry and the block symmetry. We note that equation (6) in this form only holds for points $\mathbf{r}$ inside the copper block. Analytically it is possible to derive an analogous expression for points on the boundary (see [9]). Because the latter requires the calculation of the principal value of the integral, it is numerically more actractive to use equation (6). We then calculate $T(\mathbf{r})$ with $\mathbf{r}$ in a plane inside the copper just below the DCB. Because F and thus the derivative of T normal to the DCB are known, it is straightforward to obtain a good approximation for T on the DCB.

Axial symmetry

In this case DCB is a disc . We use cylindrical coordinates (r,ϕ,z), choosing the centre of the DCB as origin and the positive z axis pointing into the copper and normal to the DCB. Due to the axial symmetry the ϕ dependence can be dropped everywhere. From equation (6) the temperature $T(r,z)$ with $z > 0$ is given by

$$T(r,z) = \frac{1}{2\pi k_0} \int_0^{2\pi} d\phi \int_0^R dr' r' (\alpha - \beta \cos \phi)^{-1/2} F(r)$$

with α and β defined by

$$\alpha = z^2 + r^2 + (r')^2$$

$$\beta = 2rr'$$

Reduction to a 1-dimensional integral is obtained by introducing the famous elliptic integral I by

$$I(\alpha,\beta) = \int_0^{2\pi} d\phi (\alpha - \beta \cos \phi)^{-1/2}, \quad \alpha > 0, \beta < \alpha.$$

For the calculation of I rapidly converging power series are available for all allowed values of α and β. The integrand of the remaining 1-dimensional integral

$$T(r,z) = \frac{1}{2\pi k_0} \int_0^R dr' \; r' I(\alpha,\beta) F(r)$$

consists of smoothly varying functions and the integration is easily performed by e.g. Simpson's rule.

Block symmetry

In this case the *DCB* is a square . We use cartesian coordinates choosing the centre of the *DCB* as origin and the positive z axis pointing into the copper and normal to the *DCB*. As pointed out above, inside the copper block T is given by the integral:

$$T(\mathbf{r}) = \frac{1}{2\pi k_0} \int_{DCB} d\mathbf{r}' \frac{1}{|\mathbf{r}-\mathbf{r}'|} F(\mathbf{r}')$$

Because of the symmetry F is invariant under the following group of operations P_i, which work on the x and y coordinates of some vector (x,y,z).

$$\begin{array}{ll} P_1(x,y,z) = (x,y,z) & P_2(x,y,z) = (-x,y,z) \\ P_3(x,y,z) = (x,-y,z) & P_4(x,y,z) = (-x,-y,z) \\ P_5(x,y,z) = (y,x,z) & P_6(x,y,z) = (-y,x,z) \\ P_7(x,y,z) = (y,-x,z) & P_8(x,y,z) = (-x,-y,z) \end{array}$$

By means of these orthogonal transformations the integration domain *DCB*, given by $\{-R \leqslant x,y \leqslant R\}$, can be reduced to an irreducible part *IP*. A possible choice for *IP* is the triangle $\{0 \leqslant x \leqslant R, 0 \leqslant y \leqslant x\}$. Using the P_i we may write

$$T(\mathbf{r}) = \frac{1}{2\pi k_0} \sum_{i=1}^{8} \int_{IP} d\mathbf{r}' \frac{1}{|\mathbf{r}-P_i(\mathbf{r}')|} F(P_i(\mathbf{r}'))$$

$$= \frac{1}{2\pi k_0} \sum_{i=1}^{8} \int_{IP} d\mathbf{r}' \frac{1}{|P_i^{-1}(\mathbf{r})-\mathbf{r}'|} F(\mathbf{r}')$$

So $T(\mathbf{r})$ is obtained by integrating over the smaller domain *IP* but with respect to the 8 vectors $P_i^{-1}(\mathbf{r})$.

We want to calculate $T(\mathbf{r})$ with $\mathbf{r}$ close to the boundary. This means that the integrand may be very steep. To deal with this kind of integrand we adopted the following approach. Let us divide the triangle *IP* into N similar smaller ones t_j. If we define

$$I_j(\mathbf{r}) = \int_{t_j} d\mathbf{r}' \frac{1}{|\mathbf{r}-\mathbf{r}'|} F(\mathbf{r}') \, ,$$

then the temperature is given by the double sum

$$T(\mathbf{r}) = \frac{1}{2\pi k_0} \sum_{j=1}^{N} \sum_{i=1}^{8} I_j(P_i^{-1}(\mathbf{r})) \, .$$

It remains to calculate $I_j(\mathbf{r})$. Because F is a smooth function we expand it linearly over each t_j. The coordinate system and the t_j can be chosen such that the bases of all t_j are along the x and y axis. Denoting the value of F at the rectangular point (x_j,y_j) of t_j by F_1 and at the other points by F_2 and F_3 the expansion of F over t_j is given by

$$F(x',y') = F_1+(x'-x_j)\frac{(F_2-F_1)}{h}+(y'-y_j)\frac{(F_3-F_1)}{h}$$

where h is the base length.
For some point $\mathbf{r} = (x,y,z=\varepsilon)$ inside the copper the integration over t_j is given by

$$I_j(x,y,\varepsilon) = \iint_{t_j} dx'dy'[\varepsilon^2+(x-x')^2+(y-y')^2]^{-1/2}F(x',y')$$

where $F(x',y')$ is by the approximation given above. So we arrive at

$$G_{1,j}(x,y,\varepsilon) = \iint_{t_j} dx'dy'[\varepsilon^2+(x-x')^2+(y-y')^2]^{-1/2}[1+\frac{x'-x_j}{h}+\frac{y'-y_j}{h}]$$

$$G_{2,j}(x,y,\varepsilon) = \iint_{t_j} dx'dy'[\varepsilon^2+(x-x')^2+(y-y')^2]^{-1/2}[\frac{x'-x_j}{h}]$$

$$G_{3,j}(x,y,\varepsilon) = \iint_{t_j} dx'dy'[\varepsilon^2+(x-x')^2+(y-y')^2]^{-1/2}[\frac{y'-y_j}{h}].$$

For small values of ε it is not attractive to evaluate the integrals numerically. Fortunately they can be evaluated analytically. For each t_j and (x,y,ε) they are of the same kind. In terms of these weight factors the I_j are given by the simple relation

$$I_j(\mathbf{r}) = \sum_{k=1}^{3} G_{k,j}(\mathbf{r})F_k$$

We shall omit here the analytical expressions. To derive them use is made of the following undefined integrals (see [6]). Denoting by Q the quadratic form

$$Q = a + bx + cx^2$$

with a,b,c some constants, it holds that

$$\int dx\ Q^{-1/2} = \frac{1}{\sqrt{c}} ln(2\sqrt{cQ}+2cx+b),\ \ c>0$$

$$\int dx\ Q^{1/2} = \frac{(2cx+b)\sqrt{Q}}{4c} + \frac{(4ac-b^2)}{8c}\int dx\ Q^{-1/2}$$

Also we need:

$$\int dx\ ln(x^2+a^2) = x\ ln(x^2+a^2)-2x+2a\ arctg\frac{x}{a}$$

5. CONCLUSIONS.

Scaling properties.

The calculated T profiles in diamond are functions of the parameters R, R_0, L and F_0, if we restrict ourselves to the axially symmetric case. We remark that on beforehand much insight can be obtained by studying the scaling properties of the problem. Looking at the diamond-copper system as a whole, we observe that T is determined by the Laplace equations (3) and the boundary conditions dealt with in section 2. All boundary conditions are homogeneous, except for those at the FIO. There it holds from equations (2), (4) and (5) that

$$k(T)\frac{\partial T}{\partial n} = k_1\frac{\partial U}{\partial n} = \frac{F_0}{(\pi R_0)^2} \tag{7}$$

The Laplace equations and the homogeneous boundary conditions are invariant for a scaling of the coordinates or T by a constant multiplicative factor. So all scaling properties can be derived from equation (7). This equation is e.g. invariant for a scaling of both the coordinates (and thus of R,R_0 and L) and F_0 by the same constant factor. This means that one of the parameters R,R_0,L and F_0 may be kept constant in the calculations.

Important Quantities.

Some quantities are of special interest. They are :

a) The maximum temperature T_{max} in the centre of the FIO. In practice this quantity may not exceed some upper limit (a usual value is $\approx$ 200), because otherwise the device, which is to be cooled, sustains damage.

b) The flux F_{DCB} across the DCB.

Results.

The results, presented in [8], lead to various conclusions, of which the most important are:

1) The results for geometries with axial and block symmetry are very similar if the cylinder and block are equal in height and volume.

2) F_{DCB} has always a sharp peak where the DCB has a sharp edge, because at these points the copper block easily absorbs the heat.

3) In geometry 1 (fig. 1) an increase of the diamond volume always leads to a reduction of T_{max}, because the diamond replaces copper. This reduction shows a saturation effect, i.e. it is hardly useful to replace more than a certain amount of copper by diamond. The temperature reduction tends to depend only on the diamond volume and not on the specific shape, because in this geometry the area of the diamond-copper boundary depends merely weakly on the shape in configurations of practical interest.

4) In geometry 2 (fig. 2) the shape of the diamond is highly important. Increasing the height of the diamond cylinder has in general a negative effect on the reduction of T_{max}, because then only extra thermal resistance is introduced. Expansion of the diamond in radius is, on the contrary, highly effective because then the *DCB* area is increased.
Geometry 2 is most effective if $R/R_0 \gg 1.$ and $L/R \approx 0.25$.

5) Comparing the effectiveness of both geometries with respect to temperature reduction we may conclude the following.
If in a system with prescribed R_0, F_0 and T_{max} values it is necessary to invest so much diamond that the conditions $R/R_0 \gg 1.$ (at least about a factor of 5) and $L/R \approx 0.25$ can be fulfilled, then the two geometries are nearly equal in effectiveness. We note that this is the case in most of the applied heat sink systems. The choice between the two possibilities in these cases has to be determined on account of other reasons (e.g. production costs).
If, however , application of a smaller diamond volume is sufficient (e.g. because only a relatively small temperature reduction is required) , then geometry 1 is clearly the more effective one.

APPENDIX.

To show some essential features of the diamond-copper heat sink system we present here an analytical solution for the 1-dimensional case.
Consider a long, thin rod along the x-axis (see *fig.* 4). For $0 \leqslant x \leqslant l_c$ it consists of diamond, for $l_c < x \leqslant l_c + l_d$ it consists of copper. At $x = 0$ a permanent heat flux F_0 is present directed into the diamond. At $x = l_c + l_d$ we assume the rod to be at room temperature and thus $T=0$. The simplicity of the 1-dimensional case lays in the fact that the heat flux must be everywhere the same. Because we only want to illustrate the general behaviour of the system we choose for convenience the values $k_0 = \alpha = 1$ and $\beta = 2$ in the expressions for the conductivity in § 2. Then the solutions of the respective heat diffusion equations are:

$$\text{in diamond: } T(x) = (F_0 x + \gamma)^{-1}$$
$$\text{in copper : } T(x) = F_0(l_c + l_d - x)$$

In *fig.* 4 these profiles are drawn.
The value of the integration constant γ must follow from the condition that at $x = l_c$ the two expressions must be equal.
We observe that this value should be positive because otherwise the temperature in diamond should increase with increasing x which is physically not admissible. So we arrive at the condition

$$\gamma = \frac{1}{F_0 l_c} - F_0 l_d > 0.$$

For given values of l_c and l_d this condition states that there exist an upper limit F_m for F_0 given by

$$F_m = (l_c l_d)^{-1/2}$$

If F_0 approaches F_m the T value at $x = 0$ tends to infinity which reflects that a stationary situation is no longer possible. Contrary to diamond, in copper always a stationary state exists thanks to its constant conductivity. In the same way it is shown that if F_0 approaches the value F_n given by

$$F_n = (l_c(l_c+l_d))^{-½}$$

the application of diamond is no longer useful. In that case only copper could rather be used.

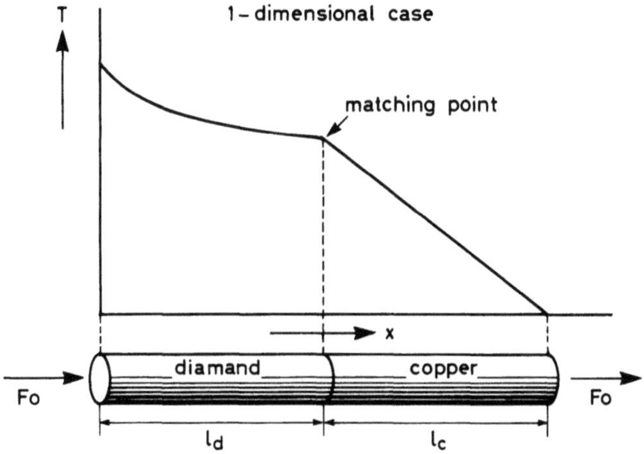

Figure 4 . Schematic picture of the temperature behaviour in a diamond/copper rod. The temperature flux F_0 in this 1-dimensional case is everywhere the same. If F_0 approaches some critical value F_m, given in the text, the temperature at x=0 tends to infinity.

ACKNOWLEDGEMENT.

This work was carried out under a contract from Drukker International B.V., Cuijk, the Netherlands.

REFERENCES.

1 Adams J.,Swarztrauber P. and Sweet R., 'FISHPACK, Library for the solution of separable partial differential equations', Boulder, U.S.A., 1980.

2 Berman R., 'Diamonds as heat sinks-a review', Diamond Research 1970, Industrial Diamond Information Bureau, London, 1970.

3 Buergemeister E.A. and Seal M., 'Thermal conductivities of diamonds with absorption at 3.22 mum', Nature, Vol. **279** , No. 5716, p.785, 1979.

4 Enckevort W.J.P. van, private communication.

5 Farlow S.J., *Partial differential equations for scientists and engineers,* J.Wiley & Sons, Chicester, England, 1982.

6 Gradshteyn I.S. and Ryzhik I.M., *Tables of Integrals, Series and Products,* Academic Press, New York, 1980.

7 Mackie A.G., *Boundary Value Problems,* Oliver and Boyd, London, 1965.

8 Molenaar J., Staarink G.W.M., 'Temperature Distributions in Diamond Heat Sinks I and II', Report no WD 85-01 and WD 85-02, Wiskundige Dienstverlening, Toernooiveld, 6525 ED Nijmegen, 1985.

9 Roach G.F., *Green's Functions,* Van Nostrand Reinhold Co., London, 1965.

ACCELERATED FATIGUE-TEST WITH STOCHASTIC LOAD-FUNCTIONS FROM THE RAINFLOW-MATRIX *

Dr.-Ing. J. D. Petersen
Raiffeisenring 14
8074 Gaimersheim
Germany

ABSTRACT. It is essential that critical safety components, for example suspension components of a car, do not fail during the service life of the vehicle. On the other hand suspension components are designed for a finite life, i.e. they may be subjected to infrequent peak stresses which they will not withstand if repeated indefinitely.

Before the individual components can be fatigue tested in the laboratory with servo-hydraulic actuators, it is first necessary to define the loading for which they are to be designed. We call this loading the "reference loading" which must be available as a time function over 300,000 km for instance, in order to feed the servo-hydraulic actuators with the corresponding signals. These "reference loadings" cannot be measured as a stochastic time function over 300,000 km, so another approach is necessary.

The method used is to take short component loading measurements with strain gauges on public roads over distances of about 100 km supplemented with measurements of extreme manoeuvres (sharp acceleration/braking) and driving on extremely rough roads on factory proving grounds. The stochastic force time functions measured are subjected to the rainflow counting process (2), which provides a data reduction and eliminates the frequency content. By superimposition of the rainflow matrices for the individual measurements and extrapolation, for instance for 300,000 km, we can obtain the reference rain-flow matrix. The required force time functions (FTF) needed to control the hydraulic actuators can then be determined using the on-line reconstruction process (3).

*This paper is the first part of paper (1)

1. INTRODUCTION

Fig. 1 shows the longitudinal force measured on a passenger car for extreme loadings, with sections for very rough road operation on a simulated African desert track, and for extreme acceleration and braking manoeuvres. The counting method is applied to force time functions of this kind.

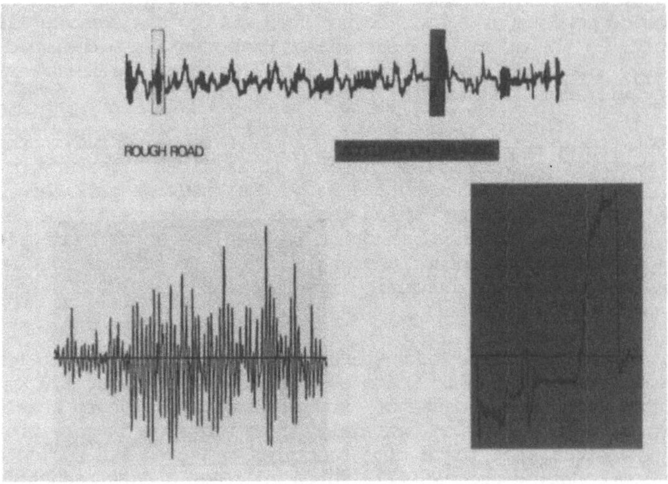

Fig. 1 : Longitudinal force

2. RAINFLOW-COUNTING

The rainflow counting method (2) counts the closed hysterisis loops in the stress/strain diagram (fig. 2). These can be seen as intermediate pulses in the FTF. The residuum left when all the intermediate pulses are eliminated can be seen in fig. 2 as the main points 1, 2 and 3.

The measurement range is divided into classes (fig. 3), and the analysis only considers the peaks, in other words the local maxima and minima. The frequency content of the measured time function is lost. In the asymmetrical rainflow a distinction is also made according to whether the intermediate pulse has been eliminated in the rising of falling part of the trace.

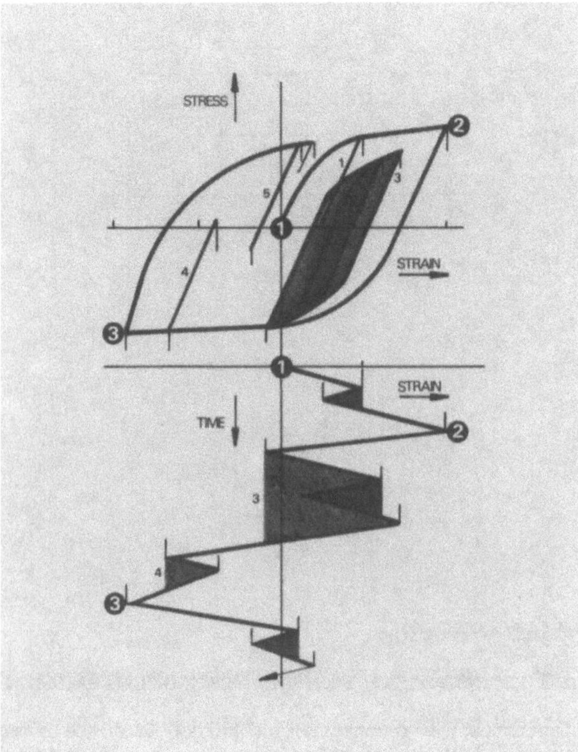

Fig. 2 : Hysterisis loop

Fig. 3 shows an example of the rainflow counting algorithm, giving an asymmetrical or symmetrical rainflow matrix as a result.

In the symmetrical rainflow no distinction is made as to whether the intermediate pulse has been eliminated from the rising or falling part of the trace. The full hysterisis loop is therefore counted symmetrically. The symmetrical rainflow matrix is obtained from the asymmetrical matrix by taking a mirror image of the elements in one triangular matrix across the diagonal and adding them to those in the other triangular matrix.

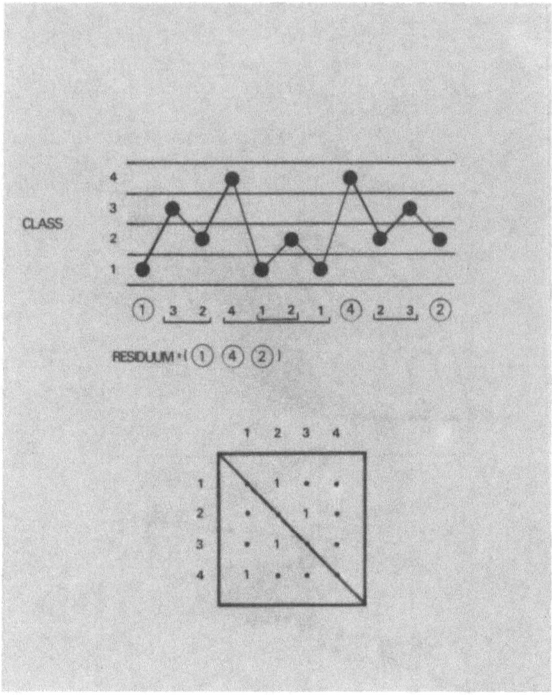

Fig. 3 : Rainflow counting

3. ON-LINE RECONSTRUCTION FROM THE ASYMMETRICAL RAINFLOW MATRIX

Fig. 4 illustrates the reconstruction method from the asymmetrical rainflow matrix. For the rainflow matrix and the residuum 2-1-4-1-3-2 there are three FTF's which have the same matrix and residuum. The on-line reconstruction method will be explained using the third FTF. The falling pulse 4-2 only fits into the residuum (1-4). Before the 4 of 4-2 can be taken out, we must check whether there are other pulses before 4-2. In this case 3-1. Let it be assumed that we have already taken out (2-1), 3-1 and 4 of 4-2. Before 2 of 4-2 can be taken out we must check whether other pulses fit into 4-2. In this case 2 x 2-3.

In other words, the on-line reconstruction process has to arrange and interpose intermediate pulses into other intermediate pulses within a prescribed residuum pair in such a way that on the one hand the rainflow matrix and the residuum are empty at the end of the reconstruction, and on the other hand the various possible decisions are taken at random with the correct probabilities. The on-line reconstruction method is new, and described in detail in (3).

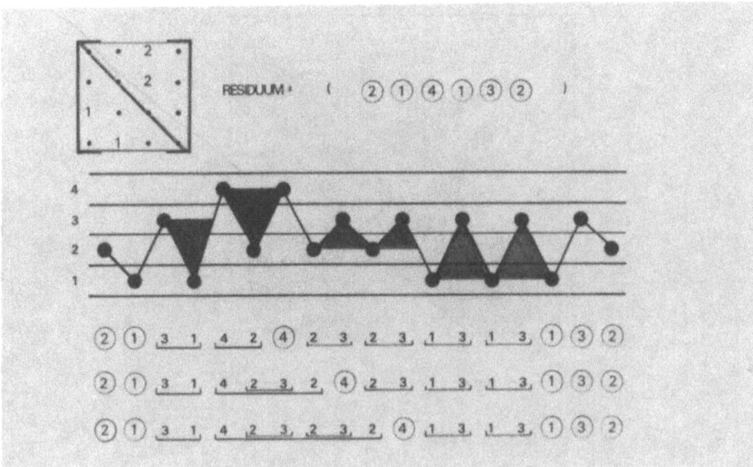

Fig. 4 : Reconstructions from rainflow matrix

The Simulation Requirement

The mathematics of the superimposition and extrapolation of rainflow matrices is explained comprehensively in (4, 5). Fig. 5 shows the simulation requirement for the longitudinal force in schematic form.

The area marked by the continuous line shows the elements that are not equal to zero from the asymmetrical rainflow matrix for 100 km driven by the customer. The area marked by the dotted line shows the extreme loading imposed by the rough road test. The line of alternative dashes and dots corresponds to extreme acceleration and braking. The continuous heavy line marks the areas resulting from extrapolation to 300,000 km. Cancelling out the adjacent diagonal elements eliminates small pulses which lie within the fatigue strength range and cause no damage. This enables us to obtain a time compression suitable for laboratory testing. The shaded area then provides the simulation requirement for the component test after application of the new on-line reconstruction method (2).

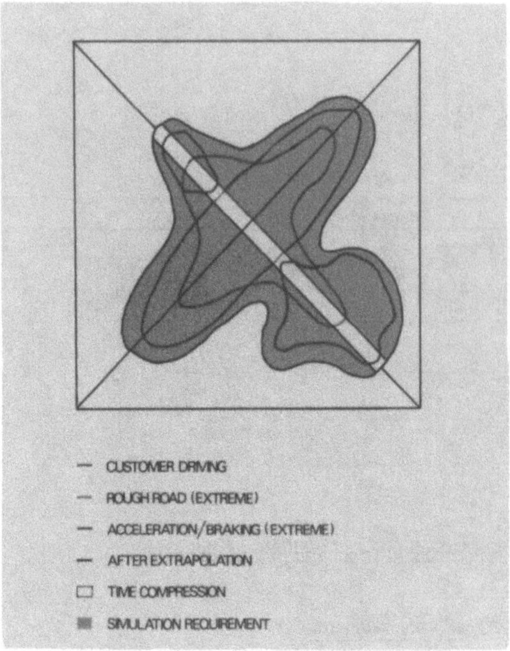

Fig. 5 : Simulation requirement

References

(1) J. D. Petersen: Evaluating Durability of Passenger Car Components: Measurement - Calculation - Simulation, Proceeding of the Workshop: The Road-Vehicle-System and Related Mathematics, B. G. Teubner Stuttgart 1985.

(2) J. B. de Jonge: The Analysis of Load-Time Histories by Means of Counting Methods, National Aerospace Laboratory NLR.

(3) Krüger, Scheutzow und Beste, Petersen: Vergleich von Markov- und Rainflow-Inversionen von stochastischen Beanspruchungszeitfunktionen. VDI-Fortschrittsbericht, Reihe 18, Nr. 22, 1985.

(4) Krüger, Petersen: Rekonstruktion von stochastischen Beanspruchungszeitverläufen aus extrapolierten Rainflow-Matrizen. 9. GESA-Symposium, Berlin, Mai 1985.

(5) W. Krüger: Mathematische Methoden zur Simulation und Extrapolation von Rainflow-Matrizen. Bericht der AGTM-Kaiserslautern Nr. 8,1985.

COMPUTATIONAL SIGNAL GEOMETRY AND LASER OPTICS

Walter Schempp
Lehrstuhl fuer Mathematik I
University of Siegen
D-5900 Siegen
Federal Republic of Germany

ABSTRACT. This paper deals with a solution of the invariant problem of computational signal geometry. As an application to laser optics, the radial intensity distribution of transverse eigenmodes inside cylindrical optical resonators and distributed-index round waveguides is explicitly computed by group theoretical methods. These calculations give rise to the transverse eigenmode patterns in the output beam of a laser as depicted in Section 3 below (cf.[1]). The calculated symmetrical field patterns transverse to the beam direction have been verified experimentally.

1. AMBIGUITY FUNCTIONS

The notions of radar/sonar auto- and cross-ambiguity functions are on the borderline with mathematics, physics, and electrical engineering. These functions have been introduced independently into analog radar signal design and analog signal processing by the engineer P.M. Woodward in his classic monograph [14] and by J. Ville [11]. If f and g denote the complex envelopes of two unitary signals belonging to the Schwartz-Bruhat space $\mathcal{S}(\underline{\underline{R}})$ on the real time line $\underline{\underline{R}}$, the corresponding analog radar cross-ambiguity function $H(f,g;.,.)$ takes the (symmetrized) form

$$H(f,g;x,y) = \int_{\underline{\underline{R}}} f(t+\tfrac{1}{2}x)\bar{g}(t-\tfrac{1}{2}x)e^{2\pi iyt}d\mu(t),$$

where (x,y) denote the coordinates of the symplectic time-frequency plane $\underline{\underline{R}} \oplus \underline{\underline{R}}$ and μ is Lebesgue measure of $\underline{\underline{R}}$. In the case f=g, the analog radar auto-ambiguity function $H(f;.,.) = H(f,f;.,.)$ of the time-varying phasor amplitude $f \in \mathcal{S}(\underline{\underline{R}})$ arises. A 'dual' version of $H(f;.,.)$ is the Wigner quasi-probability distribution function

$$P(f;q,p) = \int_{\underline{\underline{R}}} f(q+\tfrac{1}{2}t)\bar{f}(q-\tfrac{1}{2}t)e^{2\pi ipt}d\mu(t)$$

which governs the distribution of energy levels in the atomic nucleus; see Wigner [12] and the series of papers [7]. The quantum mechanical phase space distribution function P(f;.,.) and related distribution functions have been studied extensively in quantum thermodynamics since this technique provides a useful tool for the treatment of quantum problems in terms of classical concepts.

The range $\Phi = H(f;\underline{\underline{R}},\underline{\underline{R}})$ of the radar auto-ambiguity function is called to be the ambiguity surface over the symplectic time-frequency plane $\underline{\underline{R}} \oplus \underline{\underline{R}}$ generated by the phasor $f \in \mathcal{S}(\underline{\underline{R}})$. A knowledge of the geometry of Φ is particularly important since

(i) Φ represents geometrically the radar uncertainty principle and is therefore a useful computational geometric tool for analog radar signal design;

(ii) Φ is preserved in paraxial approximation as a geometric object along optical systems. The ambiguity surface conservation principle of Fourier optics determines geometrically the group of linear transformations of Gaussian beams;

(iii) Φ gives rise to the radial intensity distribution of transverse eigenmodes in cylindrical optical resonators and distributed-index round waveguides as displayed in Section 3 below. In laser optics, the transverse eigenmodes govern beam divergence, beam diameter, and energy distribution.

According to Felix Klein's Erlangen Program the group of invariants of Φ determines the geometry of Φ. The invariant problem of computational signal geometry is to calculate explicitly the group of energy preserving linear automorphisms of Φ. The present paper provides a short introduction to its solution which allows to deal with radar/sonar signal design and linear light, microwave and laser optics by computational geometric methods.

2. THE HEISENBERG GROUP

A quarter of a century ago, a thorough mathematical investigation of the radar ambiguity functions has been presented by Wilcox [13]. Unfortunately, the solution of the radar synthesis problem established by him is unsatisfactory in several respects. In 1982, however, a totally new and quite different approach to the mathematical theory of radar/sonar auto- and cross-ambiguity functions has been pointed out by the author when he first discovered the link between radar/sonar signal design and harmonic analysis on the real Heisenberg nilpotent Lie group $\tilde{A}(\underline{\underline{R}})$. By $\tilde{A}(\underline{\underline{R}})$ we understand the universal covering group of the central extension $A(\underline{\underline{R}})$ by $\underline{\underline{T}}$ of the additive group $\underline{\underline{R}} \oplus \underline{\underline{R}}$. The multiplication law of $A(\underline{\underline{R}})$ is given by

$$(x_1,y_1,\zeta_1).(x_2,y_2,\zeta_2) = (x_1+x_2,y_1+y_2,\zeta_1\zeta_2\, e^{2\pi i x_1 y_2})$$

which gives rise to the dual pairing presentation of $\tilde{A}(\mathbb{R})$

$$(x_1,y_1,z_1).(x_2,y_2,z_2) = (x_1+x_2,y_1+y_2,z_1+z_2+x_1y_2)$$

and the realization of $\tilde{A}(\mathbb{R})$ by the unipotent matrices

$$\begin{pmatrix} 1 & x & z \\ 0 & 1 & y \\ 0 & 0 & 1 \end{pmatrix} = (x,y,z) \in \underline{SL}(3,\mathbb{R}).$$

The real Lie algebra of the connected and simply connected Lie group $\tilde{A}(\mathbb{R})$ defines the canonical commutation relations of non-relativistic quantum mechanics whence the name. See the author's series of papers [2-8]and the monograph [9] on these topics. The group theoretical approach makes the quantum mechanical arguments by Ville [11] more precise and covers completely the theory due to Wilcox [13]. What is more, the author's papers referred to above show that a solution of the invariant problem of computational signal geometry may be achieved via the Segal-Shale-Weil metaplectic (or linear oscillator) representation

$$\underline{Mp}(1,\mathbb{R}) \ni \tilde{\sigma} \rightsquigarrow T_{\tilde{\sigma}} \in \underline{U}(L^2(\mathbb{R}))$$

of the three-dimensional real metaplectic group $\underline{Mp}(1,\mathbb{R})$. Indeed, the following solution of the invariant problem holds.

Theorem 1. Let the unitary operator

$$S: L^2(\mathbb{R}) \longrightarrow L^2(\mathbb{R})$$

be a linear automorphism of the radar auto-ambiguity surface over the symplectic time-frequency plane $\mathbb{R} \oplus \mathbb{R}$ - then there exists a unique transformation $\sigma \in \underline{Sp}(1,\mathbb{R})$ and a complex number $\zeta_{\tilde{\sigma}}$ of modulus $|\zeta_{\tilde{\sigma}}| = 1$ such that

$$S = \zeta_{\tilde{\sigma}} T_{\tilde{\sigma}}$$

holds.

The preceding theorem gives rise to a procedure for calculating the energy-preserving linear automorphisms of any given ambiguity surface Φ over the symplectic time-frequency plane $\mathbb{R} \oplus \mathbb{R}$ by means of chirp waveforms (linear frequency modulated signals). Since it restricts nicely to the pseudo-orthogonal subgroups of $\underline{Mp}(1,\mathbb{R})$ acting on $\mathbb{R} \oplus \mathbb{R}$, the theory of reductive dual pairs [10] allows to deal with the particularly important case of radial ambiguity surfaces over the symplectic time-frequency plane $\mathbb{R} \oplus \mathbb{R}$ in a very efficient way. Considered as a projective linear representation of the symplectic group $\underline{Sp}(1,\mathbb{R})$ acting on $\mathbb{R} \oplus \mathbb{R}$, the 2-cocycle of the metaplectic representation determines the Gouy phase-shift factor via the group theoretical Leray-Maslov-Souriau index. The family of Φ-generating chirp phasors displayed below illustrates geometrically the phase discontinuity caused by the Gouy phase-shift in optical systems.

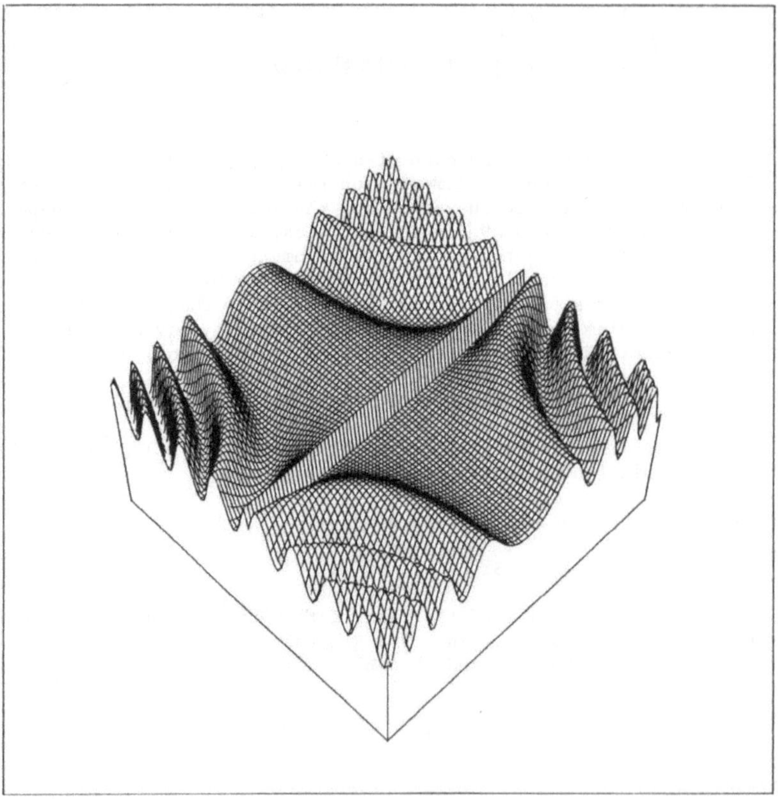

3. CIRCULAR SYMMETRY

As an application of Theorem 1 supra we get the following result:

Theorem 2. The radar auto-ambiguity surface Φ over the symplectic time-frequency plane $\underline{\underline{R}} \oplus \underline{\underline{R}}$ is circularly symmetric if and only if

$$\Phi = H(\zeta h_n;\underline{\underline{R}},\underline{\underline{R}})$$

where $\zeta \in \mathbb{C}$ and h_n denotes the Hermite function of degree $n \geq 0$.

The figure below shows a typical example of a circularly symmetric radar auto-ambiguity surface computed by the preceding theorem.

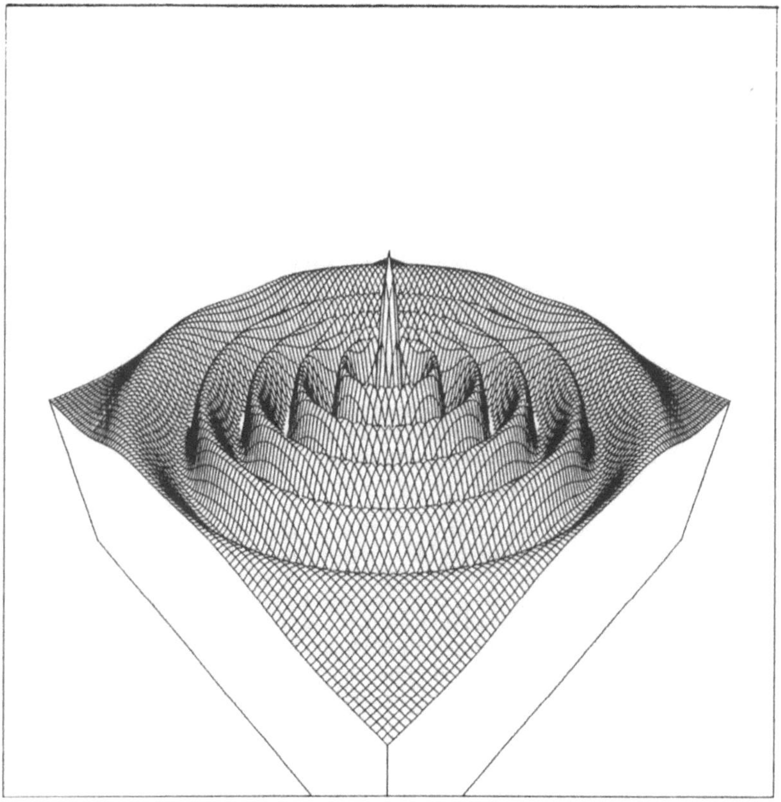

Corollary. For all integers $m \geq n \geq 0$ we have

$$H(h_m, h_n; x, y) = \sqrt{\frac{n!}{m!}} (\sqrt{\pi}(x+iy))^{m-n} L_n^{(m-n)}(\pi(x^2+y^2))$$

where $(x,y) \in \mathbb{R} \oplus \mathbb{R}$ and $L_n^{(\alpha)}$ denotes the Laguerre function of degree $n \geq 0$ and order $\alpha \geq 0$.

In laser optics, the quantity

$$|\mathrm{Re}\, H(h_m, h_n; r, \theta)|^2$$

is the radial intensity distribution of transverse eigenmodes inside cylindrical optical resonators and distributed-index round waveguides with polar coordinates (r,θ) in a plane transverse to the beam direction. The figure displayed below shows typical transverse eigenmode patterns in the output beam of a laser calculated by this group theoretical approach. Although the actions of the real Heisenberg nilpotent Lie group $\tilde{A}(\mathbb{R})$ as well as the real metaplectic group $\mathrm{Mp}(1,\mathbb{R})$ are by no means evident from the geometrical point of view, the group representational solution of the invariant problem of computational signal geometry yields highly symmetrical laser beam patterns with a characteristic number of circumferential and nodal lines. These circularly symmetric forms which have been demonstrated in special experiments using internal laser mirrors and very careful mirror alignment have exactly the same appearance, except for changes in size, at every transverse plane both inside and outside the laser ("ambiguity surface conservation principle"). There is no distinction between near-field and far-field patterns, or diffraction patterns, in the circularly symmetric case. - Note that most real laser devices can, and generally do, oscillate in any of the eigenmodes. Finding appropriate transverse-mode-control techniques so as to obtain laser oscillation only in a single controlled transverse eigenmode is one of the major practical problems in laser design.

If an eigenmode emerging from a laser oscillator is injected into a distributed-index round waveguide, a set of eigenmodes of the waveguide is excited. The coupling of the various transverse eigenmodes depends upon the incoming eigenmode, the parameters of the optical waveguide, and the degree of optical alignment. It is the purpose of a forthcoming paper to compute the coupling and conversion coefficients of the power transfer by means of group theoretic methods.

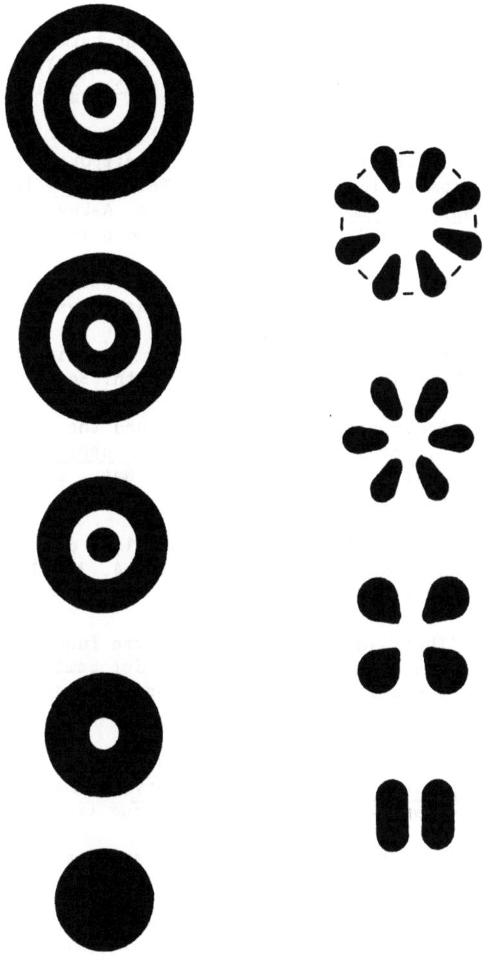

REFERENCES

1. Hebsaker, H.-M., Schempp, W.: 'Graphische Darstellung räumlicher Objekte mit Anwendungen in der Radar-Ortung'. Elektronische Rechenanlagen **25**, 32-38 (1983)

2. Schempp, W.: 'Radar reception and nilpotent harmonic analysis I-VI'. C.R. Math. Rep. Acad. Sci. Canada **4**, 43-48 (1982); **4**, 139-144 (1982); **4**, 219-224 (1982); **4**, 287-292 (1982); **5**, 121-126 (1983); **6**, 179-182 (1984)

3. Schempp, W.: 'Radar ambiguity functions, nilpotent harmonic analysis, and holomorphic theta series'. In: Special functions: Group theoretical aspects and applications. R.A. Askey, T.H. Koornwinder, and W. Schempp, eds. Dordrecht-Boston-London 1984

4. Schempp, W.: 'Radar ambiguity functions of positive type'. In: General inequalities **4**. W. Walter, ed. Basel-Boston-Stuttgart 1984

5. Schempp, W.: 'Radar ambiguity functions, the Heisenberg group, and holomorphic theta series'. Proc. Amer. Math. Soc. **92**, 103-110 (1984)

6. Schempp, W.: 'Radar ambiguity functions and the linear Schrödinger representation'. In: Anniversary volume on approximation theory and functional analysis. P.L. Butzer, B. Sz.-Nagy, and R.L. Stens, eds. Basel-Boston-Stuttgart 1984

7. Schempp, W.: 'On the Wigner quasi-probability distribution function I-III'. C.R. Math. Rep. Acad. Sci. Canada **4**, 353-358 (1982); **5**, 3-8 (1983), **5**, 35-40 (1983)

8. Schempp, W.: 'Theta identities via Laguerre functions - a nilpotent harmonic analysis approach'. Rendiconti del Seminario Matematico dell'Università e del Politecnico di Torino. Fascicolo Speciale. Torino 1985

9. Schempp, W.: Harmonic analysis on the Heisenberg nilpotent Lie group, with applications to signal theory. Pitman Advanced Publishing Program. Boston-London-Melbourne (in print)

10. Schempp, W.: 'Reductive dual pairs with applications to laser resonators' (to appear)

11. Ville, J.: 'Théorie et applications de la notion de signal analytique'. Câbles et Transmissions **2**, 61-74 (1948)

12. Wigner, E.P.: 'Quantum-mechanical distribution functions revisited'. In: Perspectives in quantum theory. W. Yourgrau, A. van der Merwe, eds. New York 1979

13. Wilcox, C.H.: The synthesis problem for radar ambiguity functions. MRC Tech. Summary Report, no. **157**. Madison, Wisconsin 1960

14. Woodward, P.M.: Probability and information theory, with applications to radar. London-New York 1953

Some Industrial Problems in Dynamic Simulation

Gustaf Söderlind
Department of Numerical Analysis and Computing Science
Royal Institute of Technology, S-100 44 Stockholm, Sweden
and
Swedish Institute of Applied Mathematics
PO Box 26300, S-100 41 Stockholm, Sweden

Abstract. In this paper we present a couple of industrial problems in dynamic simulation that have been investigated at the Swedish Institute of Applied Mathematics. Both of the applications came from process industry and were closely related to problems studied in an active research project at the Royal Institute. The problems we discuss here range from special aspects of the numerical methods used to solve the differential equations, to aspects of mathematical modelling in connection with systems having modular structure.

1. Introduction

Dynamic simulation has long been an active area in applied mathematics. The applications encountered in this field come from many sources, e.g. process industry, electrical networks, rigid-body mechanics, control systems, chemical reaction kinetics etc. Each application has its own special difficulties, ranging from how to model certain phenomena mathematically, down to how one should solve the resulting system of equations numerically. However, the capability of simulation software is rapidly growing, although often at the price of specializing in one area of application only. Thus, for some applications, especially rigid-body mechanics and electrical circuits, sophisticated software is now available. The new feature is that the user does not only get a tool for solving simulation tasks, but a piece of software that also supports the mathematical modelling of the system the user wishes to study. In other areas, with a less pronounced mathematical formalism, a similar flexibility is now being pursued.

In spite of this rapid development, it seems that there are always problems that are too hard for the software. These difficulties may stem from mathematical or numerical properties of the model, but sometimes the software design itself may turn out to be a Procrustean bed - the hardest problems are those which cannot be cast in "standard" form.

In this paper we shall discuss two problems giving rise to these difficulties. The first is a numerical problem in connection with a power plant simulation, and the second is a study of methods and needs of the modern process simulation environment.

2. The steam line problem

The background of the steam line problem was that a complex mathematical model had been developed for simulating the dynamics of a large power plant. The purpose of the model was plant operator training. The model was developed for the simulation system SIMNON, [3], implemented on a VAX 11/750. Naturally, a fast response, preferably real-time simulation, was needed.

The model had a distinct modular structure, and was essentially put together from mathematical submodels of the components. The size of the model was of the order of a hundred states. However, already in the benchmark tests of the model it turned out that execution time was unacceptable. The problem contained time scales differing by several orders of magnitude and was apparently stiff during most of the integration. Although SIMNON does contain a stiff integrator of Gear type, it was believed that the difficulties were due to a stiff submodel of a 120 m long steam line in which mass transport and pressure were modelled by a nonlinear hyperbolic partial differential equation. This submodel had been transformed to a set of ordinary differential equations by a space-discretization: the steam line was subdivided into six 20 m segments, and finite differences were used to approximate the space derivatives.

2.1 Model of the steam line

The PDE model of mass transport w and pressure p in the steam line is given by the nonlinear hyperbolic system

$$\begin{cases} w_t = \alpha p_x - \beta k_f w|w| \\ p_t = \gamma w_x. \end{cases} \tag{1}$$

Here α, β and γ are proportionality constants. The nonlinear term represents friction. The friction coefficient k_f is very small compared to other quantities of interest, and is essentially determined by the roughness of the steel surface of the pipe.
The discretization of the PDE can be illustrated as follows:

mass transport w:		y_1	y_3	y_5	y_7	y_9	$y_{11} = w_{out}$
pressure p:	$p_{in} = y_0$	y_2	y_4	y_6	y_8	y_{10}	

With the notation used above, we obtain a system of ten ODE's by approximating the space derivatives with central differences:

$$\begin{cases} \dot{y}_{2j-1} = (y_{2j-2} - y_{2j} - \kappa_f y_{2j-1}|y_{2j-1}|)/\tau_1 \\ \dot{y}_{2j} = (y_{2j-1} - y_{2j+1})/\tau_2. \end{cases} \tag{2}$$

Now, if the friction term is neglected, one obtains a linear tridiagonal system $\dot{Y} = AY$. The matrix A has a very special structure; it can be brought to skew-symmetric tridiagonal form by applying a diagonal similarity transformation. Thus one can show that the eigenvalues of the matrix A are

$$\lambda_m = i\omega_m = \frac{2i}{\sqrt{\tau_1 \tau_2}} \sin \frac{2m - N - 1}{2N + 2} \pi \tag{3}$$

when A is $N \times N$ (here $N = 10$). Thus the eigenvalues are purely imaginary when friction is neglected, and the maximum angular frequency is $\omega_{max} < 2/\sqrt{\tau_1 \tau_2}$. Using $N = 10$ and given values for α and γ, we found that $\omega_{max} < 26.3$. Next, taking friction into account we could show, by linearization, that the eigenvalues of the Jacobian would have the same moduli as before, but now having (the same) real part oscillating around -1.6. Thus the eigenvalues are always close to the imaginary axis, and transients in the system damp out in approximately five seconds. This information turned out to be of prime importance to improve the numerical performance.

2.2 Numerical instability

The numerical method that was used to integrate the system was the three-step third order backward differentiation method (BDF or Gear method). The execution time was too large, and the stepsize was apparently being limited by the properties of the problem, since there was no significant increase after the transients had damped out. A number of test runs were performed, with smoothly varying forcing functions. Indeed, it turned out that the performance of the BDF3 method was just about the same as that of the classical fourth order Runge-Kutta method (RK4) which is not intended for stiff problems.

Now, there are basically two factors limiting the stepsize in an integration code with automatic stepsize selection. First, the *accuracy requirement* has to be met; the stepsize h must be short enough in order that the solution of the differential system be represented accurately. Second, the *stability requirement* must be satisfied. For a linear system with eigenvalues λ_m this means that $|h\lambda_m|$ must remain inside the stability region of the method. It is important to note that while the stability requirement remains the same throughout the entire simulation, the accuracy requirement does not. Thus, in the transient phase, the accuracy requirement typically requires that $|h\lambda_m|$ be small, often of the order of 0.1 or less. For most methods used in practice this automatically implies stability. However, during the "steady-state" phase, the stepsize should be related to the local smoothness of the solution, and this is in general not related to the spectrum but rather to properties of the forcing function. Stiffness can then be viewed as a conflict between the stability and accuracy requirements: the problem is stiff when the accuracy requirement permits the use of "large" stepsizes in the sense that $|h\lambda_m| \gg 1$ for some eigenvalue λ_m of the Jacobian matrix. Thus if a numerical integration method has a bounded stability region (e.g. RK4), then the stepsize is limited by stability considerations rather than by accuracy.

The reason that BDF3 did not work very well for the steam line equations was that BDF methods usually have poor numerical stability properties when the spectrum is close to the imaginary axis, cf. plots of the BDF stability regions in [5, pp 215-216]. In fact, the eigenvalues of our system would have to pass through the unstable region of BDF3 during a stepsize increase. A careful inspection of the test runs revealed a hardly discernible ripple superimposed on the solution curves. Thus, as soon as the stepsize would increase, the numerical stability was lost and tiny oscillations started to build up. This implied a growing local truncation error, and consequently the stepsize strategy would cut the stepsize to improve the "accuracy". Then numerical stability was recovered, and the whole process repeated itself.

2.3 Solution of the problem

To solve the problem we needed a numerical method with better stability properties, preferably an A-stable method (the stability region covers the entire negative half-plane). Such methods are known to be reliable also for nonlinear dissipative systems, [1]. Fortunately, there are A-stable BDF methods. Both BDF1, better known as the implicit or backward Euler method, and BDF2 are A-stable, cf. [5, p 215]. It was decided to try the second order BDF2 method. From the programming point of view, the change was embarrassingly simple. All that was needed was to change *one single line* of Fortran code: the statement "K=3" was replaced by "K=2".

The results were positive. Despite the fact that the order was lowered, the accuracy remained the same, and the ripple due to the previous instability was gone. In a representative transient simulation of ten seconds of real time, the execution time dropped from 7.89 seconds for the BDF3 to 2.64 seconds for the BDF2 running on a PR1ME 550 computer. For an extended simulation of the same problem, 100 seconds real time with a slowly varying steady state, the results were even better with an execution time of 2.90 seconds for the BDF2. Thus the latter method was vastly superior in stiff intervals. A larger set of test problems confirmed this behaviour, and a speed-up by a factor of 10 - 20 was typical.

However, this is not the end of the story. By changing integration methods, the steam line model was no longer a problem. One bottleneck was thus removed, but others remain, and the speed-up for the complete plant model is significantly lower. This indicates that more work has to be done, in particular in the area of how to partition the equations properly from the mathematical/numerical point of view, and how to take advantage of the structure.

3. Modular simulation

The second problem we would like to discuss is the development of simulation tools for large systems of modular structure. While certain aspects of this problem, particularly in regard to the design of numerical methods, are presently being studied at the Royal Institute of Technology we have also investigated the special needs for a modern process simulation environment at the Swedish Institute of Applied Mathematics.

The rapid development in software design and hardware (in particular the new graphical workstations) has led to a desire to develop powerful and user-friendly simulation environments. Among some important areas that can be improved in next-generation software are man-machine communication, flexibility in the formulation of simulation problems, modelling support and efficient numerical solution.

In some areas of application many of these goals have already been achieved. Two such applications are rigid-body mechanics and electrical networks, and here a development towards an expert system philosphy is also in the offing. The main reason that these areas are more developed than others is that there is a relatively simple and distinctive structure of the formalism underlying the mathematical models. Thus, these areas lend themselves better to an algorithmic approach to modelling.

In other fields, such as process simulation, the submodels represent a much larger variety of objects and hence are more diversified. Often there is little, if any, particular structure in the way mathematical models are developed. Traditionally, such models have been developed "by hand" - a quite arduous task for large-scale systems. However, even in this area we now have a development towards automated mathematical modelling. The most straightforward approach to this problem is a systematic use of input/output modelling. As we shall see, however, this form of "computer-aided mathematical modelling" leads to problems of its own.

3.1 Modular input/output modelling

In state-space modular input/output modelling, a *module* is represented by a *submodel* of the form

$$\begin{cases} \dot{x}_i = f_i(x_i, u_i) \\ y_i = g_i(x_i, u_i). \end{cases} \tag{4}$$

Here x_i is the *state* of the i:th module, u_i is the *input* and y_i is the *output* of the module. If there is a total of m modules, the *couplings* are represented by equations of the form

$$u_i = h_i(y_1, y_2, \ldots, y_m). \tag{5}$$

Hence the coupling equations specify how inputs are computed from outputs. An algorithm for modular simulation of systems of this structure can be found in [2].

In the following we shall limit ourselves to discussing linear constant coefficient systems. This is not a restriction, but simplifies the discussion of solvability questions. Thus, the total system of equations is

$$\begin{cases} \dot{x} = Ax + Bu \\ y = Cx + Du \\ u = Ey. \end{cases} \tag{6}$$

Since the system is modular, the matrices A, B, C and D are block diagonal, in which each diagonal block corresponds to a particular module. Eliminating the output y, we obtain

$$\begin{cases} \dot{x} = Ax + Bu \\ 0 = ECx - (I - ED)u. \end{cases} \tag{7}$$

It follows that in order to solve for the inputs, we have to require that $I - ED$ be nonsingular. An important special case when this requirement is fulfilled is if it is possible to develop submodels with $D = 0$, i.e. without direct feed-through. If that is the case, we even have an explicit expression for the inputs u in terms of the states x.

However, it is well-known that problems may arise if there are algebraic loops in the system ($D \neq 0$). In fact, it is in general not possible to use the state-space input/output model (4-5) with a fixed definition of inputs and outputs in the submodels. To illustrate this, consider a resistor modelled by Ohm's law. (Our model is not the simplest possible, but serves the purpose of showing some difficulties in input/output modelling). The resistor has two terminals; in each terminal we have a voltage and a current which is positive in the outward direction. Suppose that the model is

$$\begin{cases} i_1 = (u_2 - u_1)/R \\ i_2 = -i_1, \end{cases} \tag{8}$$

where the potentials u_1 and u_2 are interpreted as inputs and the currents i_1 and i_2 as outputs. Note that this is a "static" model, i.e. there is no dynamics in the model.

Now suppose that we want to connect two such resistors, $r1$ and $r2$. The coupling equations obtained by connecting terminal 2 of $r1$ to terminal 1 of $r2$ can then be written, using the familiar dot notation,

$$r1.i_2 = -r2.i_1 \tag{9}$$

$$r1.u_2 = r2.u_1. \tag{10}$$

The interesting consequence of this is that (9) is a relation between two output variables, and (10) is a relation between two inputs. The structure of (9-10) does not conform to that of (5), and the coupling equations do not yield relations from which the inputs can be computed. *Thus the model* (4-5) *cannot represent arbitrary large-scale systems of modular structure.* In spite of this, we may still claim that the coupling equations properly describe the interconnection. Hence the problems are due to the modelling, viz. to an improper choice of input and output variables on the module level.

Clearly, this problem can easily be resolved in our simple example, e.g. by changing inputs and outputs in the models so that the coupling equations are of the form (5). However, this is not a general solution to the problem of modelling large-scale systems. Quite on the contrary, that approach leads to further complications, since it would necessitate the use of several different models for the same component or module. To be more precise, one would have to use either the model $U = RI$ or $I = U/R$ depending on where the new component is to be inserted in the network.

3.2 Descriptor models and sorting

The problem outlined above can be summarized in the question "What is input and what is output?" The answer to this question is not unique. In fact, *one cannot decide what is input and what is output until the couplings have been completely specified.* Until then, one must be prepared to re-interpret the input/output role of the variables.

A model structure which allows such a re-interpretation is (here we still restrict ourselves to the linear case; a generalization to nonlinear models is straightforward):

$$\begin{cases} E_0\dot{x} = Ax + Bu \\ \quad y = Cx + Du \\ E_u u = Fy. \end{cases}$$

Systems of this structure are often referred to as descriptor systems. To systems theorists it is also well-known that the model can be constructed so that $D = 0$. The matrices E_0, A, B and C are block diagonal in the modular case. Note that the diagonal blocks of E_0 may be singular; this signifies that the dynamics of the corresponding module consists of mixed differential and algebraic equations. Further we note that we now have a fully implicit relation between the "input" u and the "output" y.

Eliminating the "output" variables y, this system can be written

$$\begin{cases} E_0\dot{x} = Ax + Bu \\ E_u u = E_x x, \end{cases} \tag{11}$$

where $E_x = FC$. The couplings are here represented by (11b). Note, however, that both of the matrices E_u and E_x may be singular (although with nonintersecting nullspaces) in a given system. Therefore (11b) should only be regarded as a way of computing the residual in the coupling equations.

When viewed as a set of residuals, it is evident that the coupling equations are underdetermined. We may therefore select a set of variables to be solved for from

(11b). This is referred to as an *assignment*; precisely one variable is assigned to each coupling equation. The assignment, which is non-unique, is often considered as a sorting of the equations, cf. [4]. If the assigned set consists of u variables only, it is sometimes referred to as the "causal" assignment. This corresponds to an input/output model of the form (4-5). However, in most cases the assigned set will contain both x and u variables, in which case it corresponds to a proper re-interpretation of inputs and outputs.

3.3 Numerical simulation of modular systems

We shall now describe a procedure for the numerical simulation of systems of modular structure. It should be considered as a more flexible alternative to the algorithm given in [2], since our algorithm can be used for any modular system with descriptor-type submodels. For simplicity we only outline the procedure for the backward Euler or BDF1 method applied to the linear system (11), although we present the technique in a predictor-corrector form which is appropriate also in the fully nonlinear case. Whether or not the assignment mentioned above is used, depends on whether one uses Newton or fixed-point iterations to solve the system equations.

Discretizing (11) by the backward Euler method, we obtain

$$\begin{cases} h^{-1}E_0(x_{n+1}-x_n) = Ax_{n+1} + Bu_{n+1} \\ \qquad\qquad E_u u_{n+1} = E_x x_{n+1}. \end{cases} \tag{12}$$

The vectors x_{n+1} and u_{n+1} are implicitly defined in (12), and we shall use Newton's method to solve this system. Thus, in each step we have to solve linear systems of the form

$$\begin{bmatrix} h^{-1}E_0 - A & -B \\ -E_x & E_u \end{bmatrix} \begin{bmatrix} x_{n+1} \\ u_{n+1} \end{bmatrix} = \begin{bmatrix} h^{-1}E_0 x_n \\ 0 \end{bmatrix}. \tag{13}$$

If this method is to be applied to nonlinear systems, one has to predict u_{n+1}, e.g. by polynomial extrapolation. Let this predicted value be denoted by u^*. The idea is to simulate all modules independently (i.e. without taking the effects of couplings into account) over a step of size h with the predicted u^* as "input". The result obtained is denoted by x^*. Then, in each step we have to solve for corrections δx and δu such that $x_{n+1} = x^* + \delta x$ and $u_{n+1} = u^* + \delta u$. These corrections satisfy the linear system

$$\begin{bmatrix} h^{-1}E_0 - A & -B \\ -E_x & E_u \end{bmatrix} \begin{bmatrix} \delta x \\ \delta u \end{bmatrix} = \begin{bmatrix} 0 \\ E_x x^* - E_u u^* \end{bmatrix}. \tag{14}$$

Note that the system matrix in (14) has the block LU decomposition

$$\begin{bmatrix} h^{-1}E_0 - A & 0 \\ -E_x & J_s \end{bmatrix} \begin{bmatrix} I & -(h^{-1}E_0 - A)^{-1}B \\ 0 & I \end{bmatrix} \tag{15}$$

where the "Schur complement" S is given by

$$S = E_u - E_x(h^{-1}E_0 - A)^{-1}B. \tag{16}$$

This block factorization is utilized for the solution of (14). Thus we obtain the following algorithm for the numerical simulation of modular systems of the form (11):

1. Use polynomial extrapolation to predict u_{n+1} by u^*.
2. Simulate all modules independently:
   Solve $(h^{-1}E_0 - A)x^* = h^{-1}E_0x_n + Bu^*$.
3. Evaluate coupling residual $r = E_x x^* - E_u u^*$.
4. Compute corrections δx and δu :
   Solve $S\,\delta u = r$
   Compute $\delta x = (h^{-1}E_0 - A)^{-1}B\,\delta u$.
5. Correct x and u: $x_{n+1} = x^* + \delta x, \quad u_{n+1} = u^* + \delta u$.

The algorithm given above should only be considered as an elementary technique for modular simulation. A few different versions of the algorithm have been tested numerically, and the results look promising. However, there are several different variants of this basic method that can be considered for practical use, and there are also many different ways of implementing them. We shall only mention a few important possibilities.

First, we have only outlined the technique for the backward Euler method. Of course the technique could equally well be used in conjunction with other suitable methods. Further, the efficiency of the technique may be greatly enhanced if one can use different stepsizes in the individual simulations of the modules (step **2**). However, in such a multirate implementation the correction for the coupling defect is considerably more complicated. Therefore it may be necessary to iterate the solution of (14), see steps **3** and **4**. Also, if the system to be solved is nonlinear, such an iteration may be necessary.

As for the iterative method, we mentioned the possibility of using a fixed-point type iteration. In order to use such a technique, one needs the assignment discussed in Sec. 3.2. The solution procedure is then somewhat different. One first constructs predicted values for the assigned variables at time t_{n+1}. Then one solves (12a) for the unassigned set. Corrected values for the assigned variables are then computed in (12b), and the process is repeated. The success of such a method depends on the choice of assignment, but also on the structure of the couplings. Thus the method is more likely to work if the couplings are relatively weak. The technique is also more related to the approach in [2], in which a causal assignment is assumed.

Acknowledgements. Part of this work was carried out at the Swedish Institute of Applied Mathematics (ITM), and part at the Royal Institute of Technology (KTH) under STUF contract 85-3242. The author would like to thank Mr. Lars Eriksson of ITM and Dr. Jesper Oppelstrup at KTH for stimulating discussions on the topics reported in this paper.

References

[1] Dahlquist, G., *G-stability is equivalent to A-stability*, BIT **18**, (1978), pp. 384-401.

[2] Eitelberg, E., *Modular simulation of large stiff systems*, in "Progress in Modelling and Simulation", F. Cellier (ed.), Academic Press 1982.

[3] Elmqvist, H., *SIMNON - An interactive simulation program for nonlinear systems*, Proc. of Simulation '77, Montreux, June 1977.

[4] Elmqvist, H., *A structured model language for large continuous systems*, thesis, Rept. LUTFD2/(TFRT-1015)/1-226/(1978), Dept. of Automatic Control, Lund Inst. of Tech., Sweden, 1978.

[5] Gear, C.W., *Numerical Initial Value Problems in Ordinary Differential Equations*, Prentice-Hall 1971.

ISNAS

AN INFORMATION SYSTEM FOR FLOW SIMULATION BASED ON THE NAVIER-STOKES EQUATIONS

prof.dr. C.B. Vreugdenhil
Delft Hydraulics Laboratory
P.O. Box 177
2600 MH Delft
The Netherlands

ABSTRACT. Six institutes in the Netherlands launched a project to develop a comprehensive information system for flow simulation based on the Navier-Stokes equations (ISNaS). The objective is to cover a great part of industrial applications in various technical fields. The requirements for the system are outlined together with the preliminary design. The project will be split up into tasks to be performed by the participants. The resulting information system will be available to potential users in various ways.

The basis for numerical simulation of flows is found in the Navier-Stokes equations, which have been known for almost 100 years. However, until recently, the difficulties in solving them for realistic conditions, such as occurring in industrial flows, have been too great. In the last 10 to 20 years, numerical methods have been developed to a level that numerical flow simulations are now possible for practical purposes. Many computer programs are available, most of which do not meet the requirements of design offices and technical research institutes. The present paper describes a project, the objective of which is to develop an information system for realistic flow simulations, based mainly on existing physical and numerical knowledge.

DHL
DELFT HYDRAULICS LABORATORY

ECN
ENERGY RESEARCH CENTER

MARIN
MARITIME RESEARCH INSTITUTE

NLR
NATIONAL AEROSPACE LABORATORY

THDelft

THD
DELFT UNIVERSITY OF TECHNOLOGY

THT
TWENTE UNIVERSITY OF TECHNOLOGY

The six institutes listed above cover a great part of industrial flow phenomena. DHL, ECN, MARIN and NLR are non-profit research institutes, each acting as specialized consultant in its field of application. DHL covers all problems involving water, including transport phenomena, water quality, water resources management, and, indeed, other fluids. ECN is involved with applied research on all questions of energy production, be it nuclear, fossil fuels, wind or solar energy. NLR is active in reseach on aerospace research (aircraft design, aviation, space flight). MARIN investigates ship design, ship propulsion, ship manoeuvring and offshore structures. The two universities in Delft and Enschede (Twente) have active groups in the field of computational fluid dynamics; they take care of the education of specialists for the research institutes; moreover, they have their own research contracts for various industries.

FLOWS
- **FORCES**
- **HEAT TRANSFER**
- **MASS TRANSFER**
- **CHEMICAL REACTIONS**
- . . . . . .

	DENSITY KG/M^3	VISCOSITY M^2/S
WATER	1000	10^{-6}
AIR	1.3	$1.5 * 10^{-6}$
OIL	980	10^{-4}

↓

NAVIER-STOKES

It could be questioned whether the various fields of application provide a sufficient common basis for a joint effort in software development. In fact, such different fluids as water, air and oil all can be described by the same Navier-Stokes equations; this also applies to many other fluids, provided they can be characterized as "Newtonian".

For practical purposes it is usually not the flows in which one is interested, but rather derived quantities such as forces, heat transfer, masstransfer. These features will therefore be included in the ISNaS project.

NAVIER-STOKES USED TO BE TOO COMPLICATED

SIMPLIFICATIONS

* **SHALLOW WATER**
* **POTENTIAL FLOW**
* **BOUNDARY LAYERS**
* **LOWER DIMENSION**
* ......

NAVIER-STOKES CAN NOW BE SOLVED FOR ENGINEERING PURPOSES

ISNAS

Before numerical methods could be used, the solution of the complete Navier-Stokes equations was too complicated. Therefore, simplifications were applied, which were designed for specific purposes, and therefore rather different for the various fields of application. For example, in hydraulics a very common approximation for flow in seas and rivers is the shallow-water equations. For many wave problems, a potential-flow model can often be used. For many applications in airplane and ship research, boundary layers are very important. Outside such layers, again, potential flow is often a good approximation. In general, a reduction of complexity has been sought by reducing the dimension of the problem.

Today, sufficient numerical methods are available to solve the Navier-Stokes equations, both in 2 and 3 dimensions. The ISNaS project is intendend to provide a common information system for the six institutes to solve their whole range of flow problems.

PARTICIPANTS' INTERESTS

DHL - HYDRAULIC CONSTRUCTIONS
- INDUSTRIAL FLOWS
- SEDIMENT TRANSPORT
- MANY OTHER APPLICATIONS

ECN - HEAT-EXCHANGER DESIGN
- REACTOR CONTAINMENTS

MARIN - FLOW AROUND SHIPS' STERN
- WAVE RESISTANCE
- VARIOUS OTHER APPLICATIONS

NLR - TAKE-OFF AND LANDING
- EXHAUST JETS

UNIV - FUNDAMENTAL RESEARCH
- USER SUPPORT
- EDUCATION

Some typical applications for the various institutes are listed above. The underlying criteria for the flows are usually either economy or safety. Economy can be obtained by a better design of heat exchangers such that higher process temperatures can be accepted. Ship resistance due to friction and waves obviously influences the economy of ship operations. In aircraft design, smaller wings are being used for economy; this leads to more critical conditions for safety in the take-off and landing phases. In hydraulics, the maintenance dredging of shipping channels requires careful studies of sediment transport.

Safety considerations are involved in the design of reactor containments, various industrial flows and the construction of dikes or dams to prevent flooding of populated areas.

TECHNICAL REQUIREMENTS
(SUMMARY)

. 2-D, 3-D
. COMPLICATED GEOMETRY
. TIME DEPENDENT / STEADY STATE
. COMPRESSIBLE / INCOMPRESSIBLE
. TURBULENCE
. FREE SURFACE
. HEAT TRANSFER

A closer survey of the types of application intended by the participants leads to a set of requirements for the software to be developed. Many of the flows are essentially three-dimensional, although in several cases two-dimensional or axisymmetrical flow can be a good approximation. Essential is the possibility to represent complicated geometries. This applies to aircraft, ships and many industrial circuits, but also very definitely to irregular natural forms occurring, e.g. in rivers and coastlines.

Many of the flow types mentioned before are in steady state, but transient flows should also be handled (moving parts, accelerating flows wave phenomena). This also involves the simulation of a free surface. For air and gas flows, compressibility should be taken into account.

Virtually all flow types are turbulent; therefore available turbulence models should be included. Various transfer processes (heat, moisture, suspended sediment) should be taken into account. The system should allow the user to specify source terms, transfer coefficients or even new equations; the same applies to new numerical methods. This requires a great flexibility in the structure of the system.

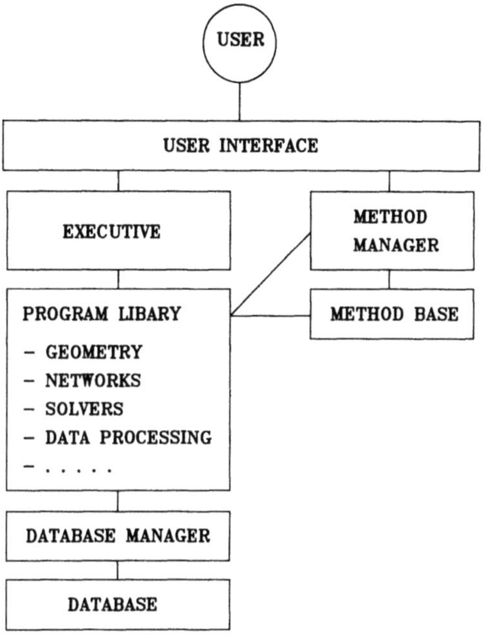

The design of the ISNaS system is outlined in the above scheme. The central idea is that the system should be so flexible that new methods can be introduced without great difficulties.
Methods will be available in a method base; they can be used as building blocks to construct programs which will perform specific tasks as indicated. A method base management system will arrange the access to the method base. Similarly, data sets will be available in a database with an efficient database management system.

The control of the entire system by the user requires a comprehensive user interface. It should be realized that there will be several types of users. Some will be interested in running standard applications; others will have non-standard problems which require careful modelling and testing. A basically different type of user is the numerical analyst who wants to introduce and test new numerical procedures. Finally, the system should be suited for educational and training purposes. It will be clear that all this requires a very careful and rather sophisticated set-up.

TASKS IN ISNAS

	PROPORTION OF COST %
1. PROJECT MANAGEMENT	7
2. STANDARDS	4
3. INTERACTION USER-SYSTEM	23
4. DATA MANGEMENT	5
5. GEOMETRY	6
6. NETWORK GENERATION	9
7. MODELS AND SOLVERS	30
8. DATA PROCESSING	7
9. DEMONSTRATION	4
10. TECHNOLOGY TRANSFER	5

The tasks in which the project is divided can be collected in three groups. Roughly one third is devoted to numerical and mathematical-physical questions (tasks 5,6,7). Another third part is related to the interaction between the user and the system, which is a key item for the performance and acceptance of the system. The remaining part is spent in various tasks of software development. Use will be made of available standard tools.

With six independent institutes as partners, a very important (though small) part of the project is to obtain agreement on the standards to be used in the software development.
To make sure that the product will be used not only by the six participants but also by other users, demonstration projects will be performed and technology transfer will be realized through courses, training, publications etc.

The total project is planned to be completed in about 5 years.

AVAILABILITY OF ISNAS

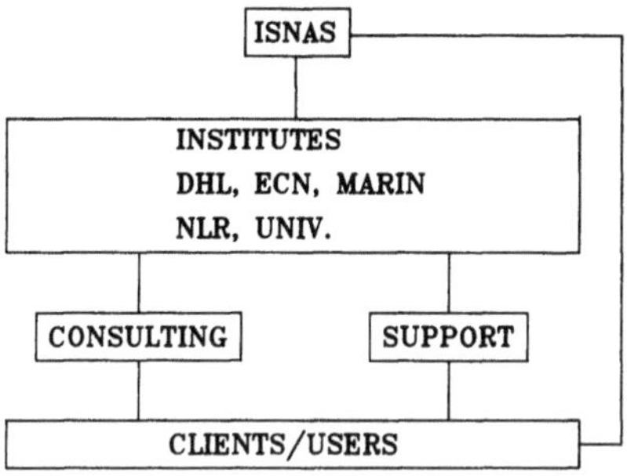

The investment to be made in ISNaS will be such that a wide use should be ascertained. The system will be available in various forms. First, the institutes involved in the development will use the system extensively for their clients in consulting projects. Secondly, the clients can get support in using the system by themselves. Finally, there will be a possibility that the software system be made available to potential users.

In all cases, it should be kept in mind that use of such a system requires a relatively high level of expertise. Even though checks and error messages will be included, a good insight into physical mechanisms and numerical methods will be required to interpret the results.

MATHEMATICAL RESEARCH AND INDUSTRIAL CONSULTING IN LINZ
POWER PLANTS - REACTION COLUMNS - HEATING OF SLABS

Hj. Wacker
Johannes-Kepler-Universität
Institut für Mathematik
Altenbergerstraße 69
A-4o4o Linz / AUSTRIA

ABSTRACT. Industrial consulting in Linz is demonstrated by three case studies. Problem I, an optimal production scheme for the OKA hydro power storage system Gosau (near the Dachstein glacier in Upper Austria) was finished in 1985. Parts of the study were already realized in practise leading to a considerable increase of production. The work was awarded with the Austrian prize for Energy Research in 1984.
The second problem concerns the determination of the thermodynamical equilibrium for distillation/absorption columns in cooperation with the VOEST. Only the first section of the work is finished yet. In the third problem, also for the VOEST, one tries to improve the process of reheating slabs in a furnace with respect to energy consumption. As work has just been started only the modelling is described. In all three cases there result highly nonlinear though well structured sparse problems. The aim is to develop problem adapted efficient solution techniques.

INTRODUCTION

Industrial consulting at the Mathematical department in Linz was started in 1975. The main motivation there was to increase the chances for our graduates when searching for a position. Further we were interested to get real world problems to develop new problem adapted and therefore, hopefully, more efficient numerical methods. Now, after a ten years experience, there is no need to help students to get positions in industry. Each student - about 2o graduates a year - is offered two and even more jobs and there exists even a kind of informal waiting list for the firms. More and more fields of applications are opened like steel industry, chemical engineering, computer science, banking, insurances, analysis of pension problems etc. Universities both in Austria and abroad are asking for our graduates working in the applied field.
Teaching was improved by installing "Problem Seminars", where real world problems from industry are to be solved by students under guidance by the staff. Since 198o courses for industry are being held.

The department now offers a 4 semester course "Mathematical Methods for Applicants".

To concentrate the national capacity a special all-Austrian research project "Method base for Applied Mathematics" was initiated by Linz in 1983 involving now 9 groups of mathematicians from 4 universities.

There is a plan to found a kind of clearing institution "Service Institute for Applied Mathematics" which could be addressed by industry for mathematical advice. Routine problems should be handled there, more difficult cases should be handed over to the department. This institution is mainly directed to the typical small firm in Austria which will not hire a mathematician but nevertheless needs mathematical advice.

We do hope that cooperation in a European frame will improve considerably the effectiveness of mathematical consulting in each country.

PROBLEM I: HYDRO ENERGY: OPTIMIZATION OF THE STORAGE PLANT SYSTEM GOSAU

Mathematical Institute: W.Bauer, H.Gfrerer, E.Lindner, A.Schwarz, Hj.Wacker
OKA (Upper Austrian Power Board): W.Barwig, .Feichtinger, R.Peßl

1. Description of the Problem

The OKA wanted us to find an optimal yearly production scheme for the storage plant system Gosau - Gosauschmied - Steeg (see Figure 1)

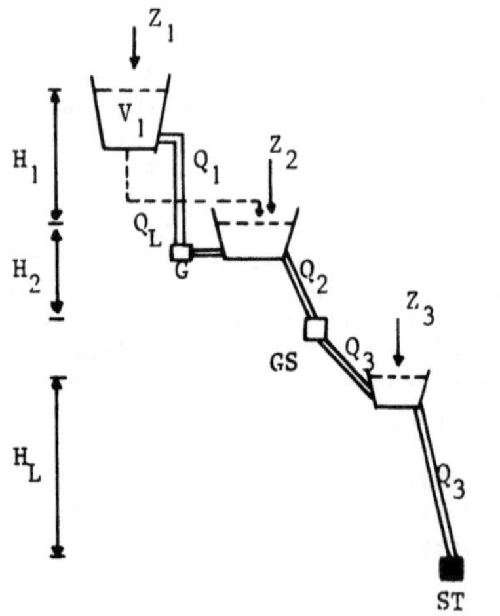

V: volume of the reservoir $[m^3]$

Q: discharges $[m^3/s]$

Z: influx $[m^3/s]$

H: head $[m]$

Q_L: seepage losses $[m^3/s]$

S: spilling water $[m^3/s]$

Optimization here means to maximize the monetary value of the yearly production respecting all constraints, technical and other ones.
The main difficulties are the following ones:

The first and largest reservoir, Lake Gosau, is situated in a carst region of the Austrian Alps and there are heavy seepage losses. Stopping production at a sea level of 92o m a.A. (meters above Adria) would empty the reservoir within a short time down to a level of 9o2.5 m a.A. The value of the production depends strongly on the actual tariff (saisonal and day/night) which is highest in winter. This contradicts the tendency given by the seepage losses. Additionally, for touristic reasons the level of Lake Gosau is to be held above 915 m a.A. during summer. Therefore, a certain amount of seepage losses has to be accepted in any case. Though plant Gosau is essential for controlling, the main part of the production is done in plant Steeg. Therefore, the whole chain must be considered when optimizing.

2. The Mathematical Model (M)

(M1) Up to a constant the objective is given by the rated energy of the system in dependence on the powers P_1, P_2, P_3.

$$E(T) = \int_o^T a(t)\ [P_1(t)+P_2(t)+P_3(t)]\ dt = \text{Max!}$$

with:

$$P_1(t) = g.\rho.f_1(V_1,V_2).Q_1.\eta_1(V_1,V_2,Q_1)$$

$$P_i(t) = g.\rho.f_i(V_i)Q_i\ \eta_i(V_i,Q_i) \qquad i = 2,3$$

Here f_i gives the head (H_i) in dependence on V_i, resp. in the case of plant Gosau, in dependence on both V_1 and V_2. The efficiency η_i generally depends on the head $H_i = f_i(V_i)$ and the discharges. As to the tariff function a(t): there is a tariff for each season (summer, intermediate, winter) and, additional, one has a day and a night tariff.

(M2) The water balance of each reservoir is given by the continuity equation:

$$\dot V_1(t) = Z_1(t)-Q_1(t)-Q_L(V_1(t))$$

$$\dot V_2(t) = Z_2(t)+Q_2(t)+Q_L(V_1(t))-Q_2(t)-S_2(V_2(t))$$

$$\dot V_3(t) = Z_3(t)+Q_2(t-\tau)+S_2(V_2(t-\tau))-Q_3(t)-S_3(V_3(t))$$

$$V_i(o) = V_{io},\ V_i(T) = V_{iT} \qquad i = 1,2,3$$

The seepage losses Q_L depend on the actual volume of Lake Gosau. Q_L is only lost for production at Gosau but adds to the influx of Gosauschmied. Spilling water is possible both for Gosauschmied and Steeg. τ is the running time from Gosauschmied to Klaushof.

(M3) $V_{imin} \leq V_i(t) \leq V_{imax}$; $Q_{imin} \leq Q_i(t) \leq Q_{imax}$ $\quad i = 1,2,3$

$0 \leq S_i(V_i(t)) \quad i = 2,3;$

$15.10^6 \leq V_1(t)$ [June 15th, October 15th];

$18.10^6 \leq V_1(t)$ [July/August]

(M4) There is a supply contract with the ÖBB (Austrian Railway Company), which has to be satisfied with highest priority
$1.5\ [MW] \leq P_3(t)$ during "day time": $t \in$ [6 a.m., 10 p.m]

3. Solution Strategy

In our case the yearly production scheme may be splitted into three independent parts. For a justification see [2].

Phase I: September 1st ($V_1 = V_{1max}$) - December 31st: (sea level: 902.5 m a.A.)
Phase II: January 1st - June 15th: (sea level: 915 m a.A.)
Phase III: June 15th - August 31st. (This problem is trivial).

Because of the small capacity both the Gosauschmied reservoir (1 day) and Klaushof (some hours) we may assume a periodicity of one day for these reservoirs.
Input data: considerable amount of work was dedicated to determine seepage losses, efficiencies and influxes.

Seepage losses: the shape of the function $Q_L(V_1)$ - see Figure 2 - was determined by analyzing the inactivity periods of more than 30 years. To get the bends exactly we used the information that some brooks in the Gosau Valley are appearing just when Lake Gosau is surpassing certain critical heights [10].

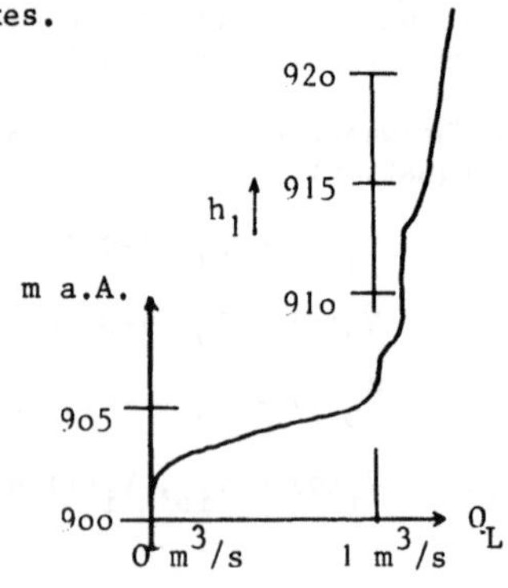

Efficiencies: the efficiency of each plant was determined by parameter identification. Based on careful measurements of the engineers a newly developed nonlinear least squares technique was used. This model also copes with the difficulty that the points where the measurements are taken are uncertain [7], [8].

Influxes: At least for Gosau the influx cannot be measured directly. The only possibility therefore is given by computation [10].

Solution of our problem was performed in the following steps:
(i) To get insight into the problem an optimal production scheme only for Gosau and Phase I was determined. Numerical solution was done by classical dynamic programming.
(ii) Next we worked out the optimal weekly production scheme for Gosauschmied. Though the reservoir Gosauschmied has only a capacity of one day we searched for a one week optimization to understand the influence of the weekend where the tariff is lower. Under some simplifications concerning the efficiency the model can be transformed by variational techniques into an equivalent finite dimensional one [5]. This model was solved then by a newly developed procedure combining a homotopy technique with an active index set strategy [3].
(iii) Solution of the full problem for Phase I including the whole chain. First a long term model using a discretization of two periods a day was solved by dynamic programming. Then, based on those results, two refined models were solved with production periods down to even one hour. Here also peak power demands were included [6], [10].
(iv) Solution of Phase II. First the optimal production scheme for each week was determined by simulation using 10 different production strategies including pumping activities. These results were refined by a quasioptimization for each week [10].
(v) Additionally, the possibilities for pumping activities were analyzed giving information to the engineers under what circumstances (e.g. V_1, V_2, price of offered pumping energy, etc.) pumping should be performed [10].

4. Numerical Solution: Decomposition/Convexification Technique

(M) is a nonlinear problem of optimal control with constraints both for the state variables (V) and the control variables (Q). After discretization there results a high dimensional nonconvex optimization problem with special structure.

We confine ourselves to describe a newly developed decomposition method applied to one of the refined models mentioned above in (iv). For ease of presentation some simplifications are made for our medium term model (1-3 weeks, discretization down to 1 hour). We assume the efficiencies to be constant. The method is described for one plant only, say Gosau, neglecting here the dependence of f_1 on V_2.

Index 1 is omitted.
To apply decomposition techniques one needs separability and convexity.

i) Separation of model (M): (MS)
After discretization and assuming the influx, the discharges and the tariff to be constant in $[t_{i-1},t_i)$ we get for the rated energy production in the interval $[t_{i-1},t_i)$:

$$E_i(V_{i-1},V_i) = a_i Q_i \int_{t_{i-1}}^{t_i} f(V_{i-1} + \frac{(V_i - V_{i-1})(t-t_i)}{\Delta t_i})\, dt$$

Separation is performed by splitting V_i: V_i^- $(t \leq t_i)$ resp. V_i^+ $(t \geq t_i)$:

(MS1) $$\sum_{i=1}^{N} E_i(V_{i-1}^+, V_i^-) = \text{Max!}$$

(MS2) $$V_i^+ = V_i^- \qquad i = 1,\dots,N-1$$

(MS3) a) $V_o^+ = V_o$, $V_N^- = V_T$ (volumes given by the long term model)

b) $V_{min} \leq V_i^+, V_i^- \leq V_{max} \qquad i = 1,\dots,N-1$

c) $0 = Q_{min} \leq \frac{V_{i-1}^+ - V_i^-}{\Delta t_i} + Z_i - Q_L(V_{i-1}^+) \leq Q_{max} \qquad i = 1,\dots,N-1$

ii) Convexification
To apply decomposition techniques one needs convexity for (MS) at least in a neighbourhood of the optimal solution. Convexity is hurt in the objective (e.g. by the tariff function) and in the Q-constraints (by the function Q_L). In our case of short time optimization convexity of Q_L is achieved by linearization with respect to V. The objective is convexified in the following way:

$$E_i \to \bar{E}_i := -E_i(V_{i-1}^+, V_i^-) + c_{i-1}^+ (Y_{i-1}^+ - V_{i-1}^+)^2 + c_i^- (Y_i^- - V_i^-)^2,$$
$$c_i^+, c_i^- > 0$$

iii) Decomposition
To solve the resulting convex problem by decomposition first the equality constraints (MS2) are included in the objective by duality:

$$\phi(Y,\lambda) := \min\{ \sum_{i=1}^{N} \bar{E}_i(V_{i-1}^+, V_i^-) + \sum_{i=1}^{N-1} \lambda_i (V_i^+ - V_i^-) \mid V \text{ subject to (MS3)}\}$$

Then we have to solve the unconstrained Minimax-problem:

$$\min_{Y \in \mathbb{R}^{2N-2}} \quad \max_{\lambda \in \mathbb{R}^{N-1}} \quad \phi(Y,\lambda)$$

To determine $\phi(Y,\lambda)$ N linearly constrained optimization problems (*) of dimension two only (!) are to be solved which can be done quite efficiently:

$$(*) \qquad \min\{\bar{E}_i(V^+_{i-1},V^-_i,Y^+_{i-1},Y^-_i)+\lambda_{i-1}V^+_{i-1}-\lambda_i V^-_i \mid V \text{ subject to (MS3)}\}$$

Algorithm: Start: $k = 0$, $Y^k \in \mathbb{R}^{2N-2}$

Step 1: Maximize $\phi(Y^k,\lambda)$ with respect to λ: λ^k

V^k: corresponding solution of (*) when evaluating (Y^k,λ^k)

Step 2: $Y^{k+1}: = V^k$, $k: = k+1$, Go to Step 1

The method is globally and superlinearly convergent, see [4]. To demonstrate the efficiency we give some information on the CPU time (one plant, accuracy w.r.t. Q: $1o^{-1o}$)

TABLE I

N	CPU-time (IBM 36o-155)	Constraints
42	14"	166
168	1'27"	67o
1oo8	35'17"	4o3o

To get a one week optimization for the whole chain using an hourly discretization one needs 14' (Q accuracy: 1 %, IBM 3o31), see [6].

5. Realization in Practise

The Results for Phase I were put into practise in 1983/84 and 1984/85. The main strategies for sinking Lake Gosau can be formulated as follows
- Start sinking exactly at October 1st (when the high winter tariff is valid)
- Don't stopp sinking until a sea level of 9o2.5 m a.A. is reached.
- Sinking is to be performed at prescribed intensities

Figure 3 demonstrates the difference between the old strategy and the optimal one (1972).

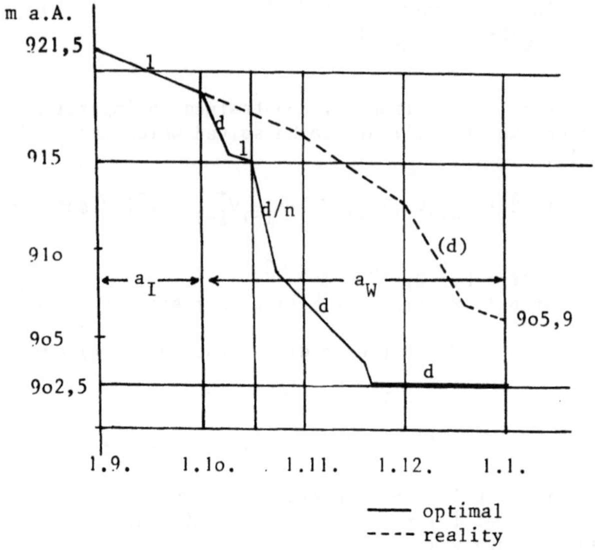

Figure 3:
1: seepage losses only
d: activity at day time only
d/n: activity during day and night

a_I: intermediate tariff
a_W: winter tariff

The following table demonstrates the efficiency of the optimization (Phase I only)

TABLE II

year	sinking time (days)	start	production GWh
1968	75	5.11.	1.75
1969	12o	2o.o9.	1.55
1971	164	1o.o8.	o.65
1974	1o9	13.11.	1.2
1979	65	15.1o.	2.o9
1982	78	22.o9.	1.93
1983	51	1.1o.	2.65
1984	45	1.1o.	2.6o

I.e.: we get an increase of more than 3o % compared to the best results from all former years. It should be pointed out that both 1983 and 1984 were dry years! An additional increase results from analyzing the seepage losses in the galleries: closing the system at non-activity periods will give another o.3 GWh a year.

For Phase II the first realization period has just started now in January 1986. Depending on the influx we await an increase of 5 % to 8 % a year. An additional increase will result from pumping activities.

References:

[1] W.Barwig/R.Peßl 'Optimization of the Gosau System' in: *Applied Optimization Techniques in Energy Problems* (Proceedings), Teubner, pp. 94-118, 1985

[2] W.Bauer/H.Gfrerer/E.Lindner/A.Schwarz/Hj.Wacker 'Optimal Production Scheme for the Gosau Hydro Power Plant System', Institutsbericht Nr. 296, Institut für Mathematik, Universität Linz, pp. 1-21, Sept. 1986 (submitted)

[3] H.Gfrerer/J.Guddat/Hj.Wacker 'A globally convergent algorithm based on imbedding and parametric optimization', *Computing* **3o**, pp. 225-252 (1983)

[4] H.Gfrerer 'Globally convergent decomposition methods for nonconvex optimization problems, *Computing* **32**, pp. 199-227 (1984)

[5] H.Gfrerer 'Optimization of Hydro Energy Storage plant problems by Variational Methods, *ZOR/B* **28**, pp. 87-1o1 (1984)

[6] H.Gfrerer 'Optimization of Storage Plant Systems by Decomposition, in: *Applied Optimization Techniques in Energy Problems* (Proceedings) Teubner, pp.215-226, 1985

[7] E.Lindner/Hj.Wacker 'Input Parameter for the Optimization of the System Partenstein: Forecasting of the Influx, Determination of the Efficiency Function' in: *Applied Optimization Techniques in Energy Problems* (Proceedings), Teubner, pp. 286-322, 1985

[8] H.Schwetlick/Tiller 'Numerical Methods for Estimating Parameters in Nonlinear Models with Errors in the Variables' Preprint Nr.69, Universität Halle, Sektion Mathematik, pp.1-25 (1982)

[9] Hj.Wacker 'Mathematical Techniques for the Optimization of Storage Plants, in: *Applied Optimization Techniques in Energy Problems* (Proceedings), Teubner, pp. 4o5-448, 1985

[1o] Hj.Wacker/W.Bauer/H.Gfrerer/E.Lindner/A.Schwarz 'Erzeugungs-optimierung in Wasserkraftwerken', Arbeitsbericht für das Bundes-ministerium für Wissenschaft und Forschung/OKA, pp.1-22o (1985)

PROBLEM II: CHEMICAL ENGINEERING: MULTISTAGE MULTICOMPONENT CHEMICAL COLUMNS

Math.Institute: D.Auzinger, L.Peer, Hj.Wacker, W.Zulehner
VOEST: F.Kokert, P.Pfeifer

1. Introduction

We consider a multicomponent mixture of liquid and volatile components. To separate the components (at least partially) one may use, e.g., distillation and absorption techniques. Wellknown examples are the distillation of alcohol or oil cracking. To increase the efficiency of such a process one uses multistage columns (see Figure 1), pumping activities, side strippers, parallel columns etc. The following model allows to determine the thermodynamic equilibrium of multistage multicomponent mixtures. The final aim is to optimize production processes including the costs of the configuration of the system and the costs for running the system.

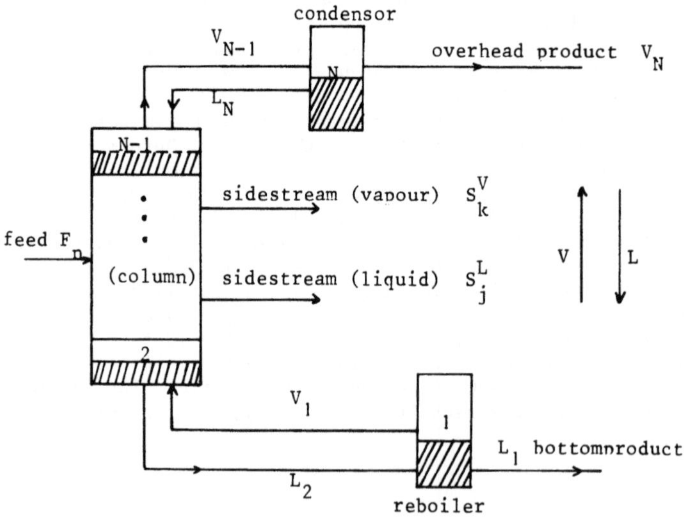

Figure 1. N-stage distillation column

In the column the vapour moves upwards, the liquid downwards. At stage N (condensor) the volatile overhead product V_N leaves the system whereas at stage 1 (reboiler) the liquid bottom product is available. Both volatile and liquid sidestreams may be extracted from the column. Further there is the possibility to insert feeds of given components and temperature.

2. The Mathematical Model

Our model describes the thermodynamic stationary equilibrium for a M-component volatile-liquid mixture for N stages simultaneously. No dynamics and no reactions are involved. To put up a mathematical model three sets of equations have to be considered for the transition between neighbouring stages (see Figure 2). Some special items for stage 1 resp. N are omitted here.

Figure 2. Transitions between neighbouring stages

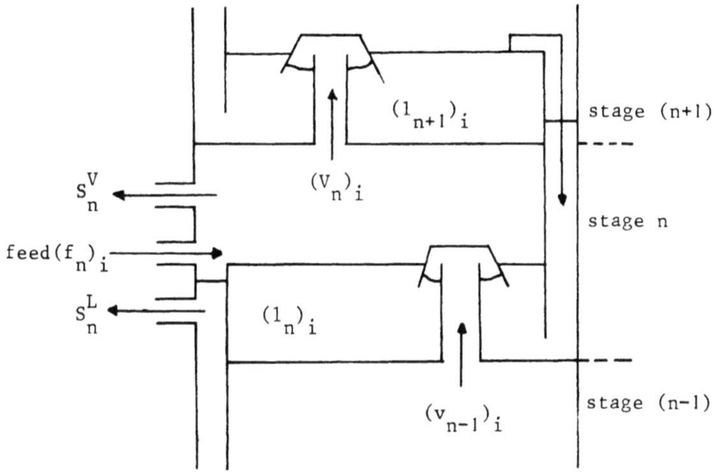

i) Component material balances: under our assumptions there holds balance for each stage n and each component i including both the liquid part of component i: $l_{n,i}$ and the volatile part of component i: $v_{n,i}$, i.e. we have equivalence between

streams entering stage n:	resp. leaving stage n:
$l_{n+1,i}$ (from stage n+1)	$l_{n,i}$ (to stage n-1)
$v_{n-1,i}$ (from stage n-1)	$v_{n,i}$ (to stage n+1)
$f_{n,i}$ (feed)	S_n^L, S_n^V (side streams)

We get for $n = 1,\ldots,N$; $i = 1,\ldots,M$ and $v_{n,i} := 0$, $l_{N+1,i} := 0$:

$$B_{n,i} \equiv \left(1 + \frac{S_n^L}{L_n}\right) l_{n,i} + \left(1 + \frac{S_n^V}{V_n}\right) v_{n,i} - l_{n+1,i} + v_{n-1,i} - f_{n,i} = 0$$

$$L_n : = \sum_{i=1}^{M} l_{n,i} \quad \text{resp.} \quad V_n : = \sum_{i=1}^{M} v_{n,i}$$

ii) Enthalpie balances: assuming the absence of any heat losses there holds energy balance for each stage n including the volatile and the liquid part of the mixture.
We get for n = 1,...,N and $V_o : = 0$ and $L_{N+1} : = 0$

$$E_n \equiv (L_n+S_n^L)h_n^L + (V_n+S_n^V)h_n^V - L_{n+1}h_{n+1}^L - V_{n-1}\, h_{n-1}^V - F_n h_n^F = 0$$

Here h_n^L, h_n^V, h_n^F are the specific enthalpie [cal/Mol] of the liquid resp. vapour resp. feed at stage n. The enthalpie depends on the components of the mixture, the temperature T_n and the (given) pressure.

In our model we use

$$h_n^V \equiv \sum_{i=1}^{M} (c_i+d_iT_n)\, \frac{v_{n_i}}{V_n} \quad \text{resp.} \quad h_n^L \equiv \sum_{i=1}^{M} (a_i+b_iT_n) + \underbrace{h_E(l_{n,1},\dots,l_{nM};T_n)}_{\text{excessenthalpie}}$$

iii) Liquid-vapour equilibrium: the thermodynamic equilibrium for each stage n and each component i can be described in the following way (n = 1,...,N; i = 1,...,M):

$$G_{n,i} \equiv k_{n,i} l_{n,i} - \frac{L_n}{V_n} v_{n,i} = 0$$

For the equilibrium constant $k_{n,i}$ we use the UNIQUAC model:

$$k_{n,i} \equiv \frac{par_{n,i}(T_n)}{p_n} + \frac{\gamma_i(l_{n,i},\dots,l_{n,M};T_n)}{\varphi_i(v_{n,i},\dots,v_{n,M};T_n)}$$

$par_{n,i}(T_n)$: partial pressure of component i

p_n: pressure of stage n

γ_i: activity coefficient

φ_i: fugacitivity coefficient

Summing up we get a highly nonlinear problem of dimension N(2M+1) for the variables $(v_{n,i}, l_{n,i}, T_n)$ n = 1,...,N; i = 1,...,M.

Additionally we have to respect nonnegativity conditions for all the variables.

3. Numerical Solution

3.1. Blocktridiagonal Structure

Arranging both the variables and the equation in the same order as the stages result in a blocktridiagonal structure for the Jacobian of the system. The bandwidth of the extremely sparse outdiagonal blocks can be further reduced by 5o % using the arrangement $(l_{n1},\ldots l_{nM};T_n;v_{n1},\ldots v_{nM})$.

For the equations we use the order $(B_{n1},\ldots,B_{nM};T_n;G_{n1},\ldots,G_{nM})$.

Figure 3 describes the structure of the Jacobian.

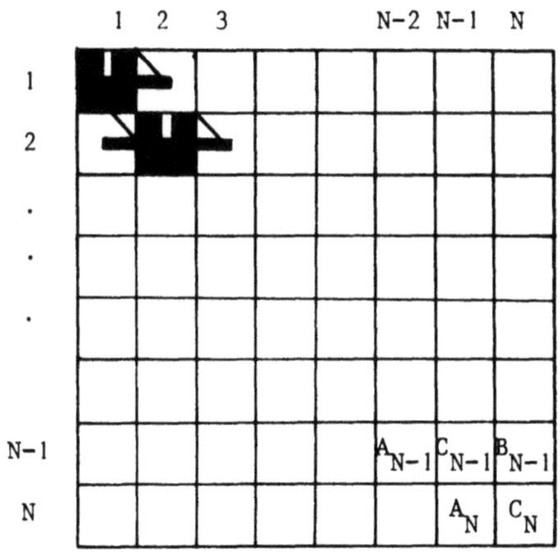

3.2. Numerical Solution Techniques

We have to solve the nonlinear equational system

$$F(x) = 0, \quad x \geq 0$$

Two solution techniques were realized:

(i) Classical Newton's Method observing the blocktridiagonal structure of F' when using L-U decomposition.

(ii) Quasi Newton techniques where first the Jacobian is updated and then L-U decomposition is used.

In both cases it was essential to use pivoting techniques, careful damping resp. restarting strategies, globalization by help of the continuation method and a suitable projection technique to cope with the nonnegativity conditions. The main argument for (ii) was to avoid the computational amount resulting from the (mainly numerical) differentiation of E_n and G_{ni}.

3.3. Extensions of the Model: Pumping Activities and Side strippers

Figure 4 gives a simple example from a 4 component oil cracking problem including both pumping (main column only) and sidestrippers. To control the pressue H_2O is inserted as an additional component (omitted in Figure 4). For the given input I the picture also shows the result of the separation process: output $O_1,\dots,O_4$; input (stage 4)

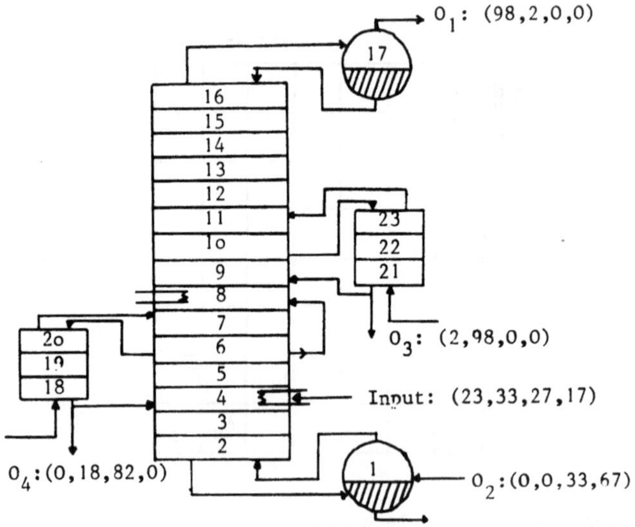

Figure 4

For each pumping activity (P) there result two extremely sparse blocks each of dimension (M+1)x(M+1) which are situated symmetrically outside the blocktridiagonal part of the Jacobian. The same situation holds when a sidestripper is added.
Again two techniques are compared

(i) Classical Newton's Method with L-U decomposition accepting that each additional pair of blocks fills a row in L resp. a column in U. The numerical amount depends, therefore, strongly on the position of those extra blocks. By a suitable pivoting technique the numerical amount for filling up the U columns can be halved.

(ii) Newton's Method is only used for the (old) blocktridiagonal system. Then the out blocktridiagonal parts are iterated off linearly.

Remark: Even for "small" systems there may be a considerable increase for the computational amount compared to the simple blocktridiagonal case. For instance, for a 4 columns problem with 15,4,3,3 stages and 5 pumping activities there may result an increase up to 2182 %!

3.4. Numerical Results

To demonstrate the efficiency of our solution techniques we present a comparison of 4 solution methods performed by Kokert on a IBM 3o81 - CPU-time in sec.
M1: Fixed point iteration (VOEST)
M2: Newton's Method (VOEST/Technische Universität Graz)
M3: Quasi Newton Technique (Auzinger [1])
M4: Newton's Method (Peer [2])

TABLE I

Example	M1	M2	M3	M4	
1. propan, butan, hexan,pentan (N=18)	82.1	3.7	3.2	1.9	
2. methanol, H_2O, ethanol (N=3o)	-	15.9	4.4	3.7	no pumping
3. ethanol, H_2O (N=3o)	5o.5	4.1	2.3	1.9	no side strippers
4. aceton, H_2O (N=3o)	6.7	1.7	1.5	1.1	
5. benzol, 2-buthan, acetonnitril (N=1o)	-	-	o.7	o.8	
6. oil cracking $R_1,..,R_n$, H_2O ($N1=17,N2=N3=3,P=1$)	not possible		*)	45" - 18o" **)	extensions

*) no comparison available
When comparing Newton's Method (i) and the Newton/Iteration technique (ii) the second method proved to be better in more than 5o % of all cases.
**)9 homotopy steps à 5" - 2o"

References:

[1] D.Auzinger 'Berechnung stationärer Zustände von Destillations- und Absorptionskolonnen (Newton - Quasi Newton Technik), diploma thesis, Mathematisches Institut, Universität Linz, 1985

[2] L.Peer 'Berechnung stationärer Zustände von Destillations- und Absorptionskolonnen (Newtontechnik/Iterationstechnik), diploma thesis, Mathematisches Institut, Universität Linz, 1985

PROBLEM III: STEEL PRODUCTION: OPTIMAL REHEATING OF SLABS IN A PUSHER TYPE FURNACE

Math.Institute: D.Auzinger, Hj.Wacker, W.Zulehner
VOEST: B.Lindorfer

1. Description of the Problem

In the last decade considerable effort has been devoted to improve the processing of reheating furnaces. For instance: hot charging (with a reduction of energy consumption up to 4o %), oxygen enrichment (with a saving of about 8 %), use of waste gases by a recuperator or by extending the convective zone at the charging end, etc. Further, one uses double stranding, increase of furnace pressure, better insulation, changing the position of skids, improvements for the burners, turning of billets etc. In our case we were consulted by the VOEST for the automatization of their pusher-type reheating furnaces. The aim was i) to optimize the reheating process with respect to energy consumption, ii) to give advice for the on-line control of the process.

Figure 1 describes schematically a 4-zones pusher type reheating furnace.

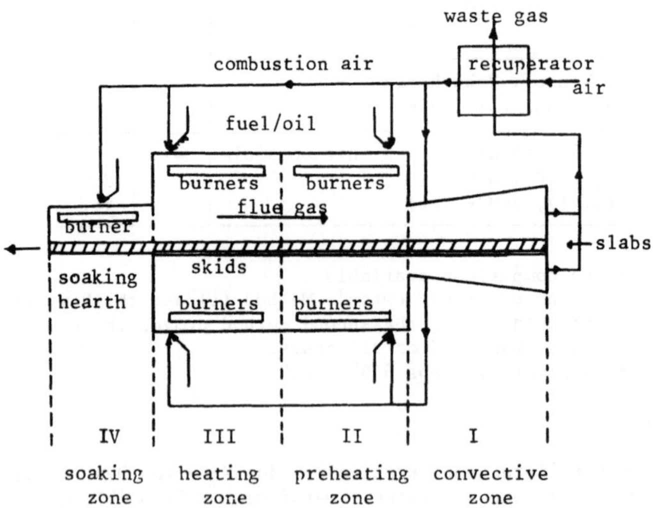

In the following we confine ourselves to cold charging for uniform slabs. The more interesting problem of mixed hot/cold charging for slabs of different types will be attacked later.

The slabs (12m x 1.2om x o.2m) enter the furnace from the right, are pushed continuously ($v \approx 12$m/h) on watercooled skids through the furnace and are discharged at a prescribed temperature necessary for the special kind of steel to be rolled immediately afterwards. Heating is performed here by combination of fuel or oil. Losses of energy are reduced by use of a recuperator where the waste gases exchange heat with the fresh air (→ combustion air). When modelling we have to take into consideration all kinds of heat-transports: convection and conduction (all zones), and radiation (II, III both top and bottom, IV top). In zone IV (bottom) there are no skids but the slabs are being moved on the soaking hearth to get an (almost) uniform temperature.

It was decided to handle the problem in a "problem seminar". The following tasks were spezified:

task 1: Solution of the heat equation for the slabs at a given temperatur of the furnace resp. flue gas.

task 2: On-line control of the furnace temperature: the flue gas temperatur is controlled in such a way that the "core" temperature of the slabs approximates a given nominal value assumed by the engineers to be "optimal".

task 3: Determination of the optimal nominal value of task 2: respecting all technical and other restrictions combustion is controlled in such a way that the slabs reach the prescribed discharging temperature with a given accuracy with minimal losses of energy.

2. The Mathematical Models (Zulehner), Numerical Techniques

Under our assumptions all models are stationary in some sense. We may use a furnace fixed coordinate system (x,y,z) describing the length, height, broadth position of a point in the slabs, resp. a slab fixed system (t,y,z) with $t := x/v$. With respect to task 2 the efficiency of the applied numerical method is most important. Nevertheless also three dimensional models are studied to understand the influence of the skids (skid-shadows) and for comparison when using models of dimension two or even one.

Ad task 1: Balancing the heat fluxes we arrive at the following three dimensional model (M1)

$$\rho(T)c(T)\,\frac{\partial T}{\partial t} = \operatorname{div}(\lambda(T)\operatorname{grad} T), \quad T = T(t,x,y) \tag{1}$$

T: temperature of the slab, ρ: density of steel, c: specific heat, λ: thermal conductivity

$$T(o,y,z) = T_{Ch} \qquad (T_{Ch}\text{: Charging temperature}) \tag{2}$$

(3)

$$\lambda(T)\,\frac{\partial T}{\partial n} = \begin{cases} \begin{bmatrix} \alpha \\ \alpha+\alpha_r \end{bmatrix} (T_f - T) & \begin{matrix} I_t \text{ (zone I top)} \\ II_t,\ III_t,\ IV_t \end{matrix} \\[2ex] (1-\frac{s}{l}) \begin{bmatrix} \alpha \\ \alpha+\alpha_r \\ 0 \end{bmatrix} (T_f - T) - \begin{bmatrix} \frac{s}{l} k_{SK}(T-T_{SK}) \\ \frac{s}{l} k_{SK}(T-T_{SK}) \\ k_{SH}(T-T_{SH}) \end{bmatrix} & \begin{matrix} I_b \\ II_b, III_b \\ IV_b \end{matrix} \end{cases}$$

α resp. α_r (T_f, T): coefficient of heat transfer resp. radiation;

l resp. s: breadth of slabs resp. skids; k_{SK} resp. k_{SH}: coefficients of heat transmission (skids/slabs) resp. (soaking hearth/slabs);

T_f, T_{SK}, T_{SH} temperature of furnace (flue gas) resp. kids resp.soaking hearth.

For case of simplicity we assume the coefficients α, k_{SK}, k_{SH} to be constant in each zone.

Numerical solution: First (M1) was solved by the Multigrid Method. Now we are working to combine the MGM with the FE technique. The latter is used to cope with the skids at a reasonable numerical amount.

Assuming $\left.\frac{\partial T}{\partial z}\right|_{\partial\Omega} = 0$ we may reduce the problem considerably by symmetries (a), (b), (c).

Figure 2. Crosscut of the slab

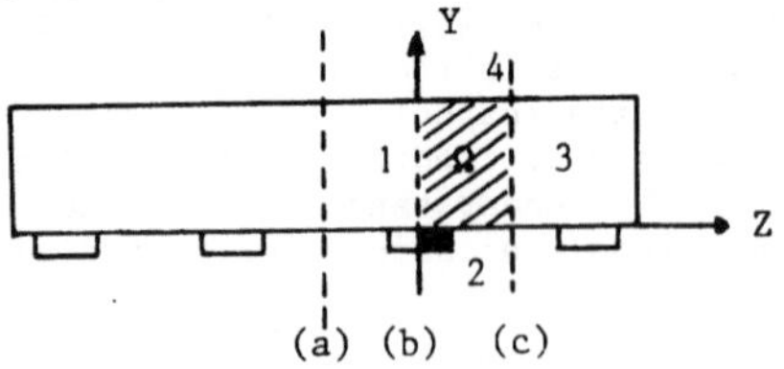

Ad task 2: Choosing to approximate the given nominal heat distribution T_N in the L_2-norm we get the following model (M2)

(1) $\int_0^{t_E} [T_N(t) - T(t,\bar{y},\bar{z})]^2\, dt = \text{Min!}$ here: $T = T(t,y,z)$

T_N: nominal heat distribution at the "core" $(\bar{y},\bar{z})$ of the slab.

(2) see (M1) (2) (3)

(3) $T_{min}(t) \leq T(t,y,z) \leq T_{max}(t) \wedge T_{fmin}(t) \leq T_f(t,y,z) \leq T_{fmax}(t)$,

$(y,z) \in \partial\Omega$

$T_f(t,y,z)$ furnace resp. flue gas temperature at $\partial\Omega$

Practise: T_f can be determined by measurements. By experience it is possible to control T_f by changing the input of combustion material (fuel/oil).

Numerical solution: In a first simplified version the furnace-temperature is described by a constant in each of the following parts of the furnace: I, II, III (top/bottom) and IV top. Therefore, we get a 7-dimensional nonlinear constrained optimization problem. The parabolic heat equation is discretized with respect to (y,z) resulting in a highdimensional system of ODEs. For the solution of the resulting optimization problem see [1], [3].

Figure 3 describes the situation for task 2.

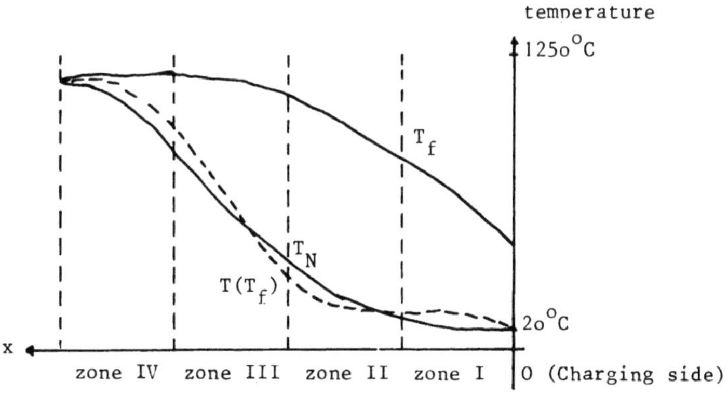

Ad task 3: We use a one dimensional model for the flue gas, splitted up into a top and a bottom part. This model is to be combined with a 1/2/3 dimensional model for the slabs. Additionally we need a model for the recuperator.

(1) A heat flow balance for the flue gas temperature $\theta(x)$ gives

$$\frac{d}{dx}\left[\dot{m}_G(x) C_M^G(\theta)\theta\right] = \left[H_u + \nu C_m^{CA}(\theta_{CA})\theta_{CA}\right]\dot{B}(x) - lk(\theta - T) - l_W k_W(\theta - T_H) \equiv (*)$$

θ, θ_{CA}, T_H:	temperature of flue gas resp. combustion air resp. working hall
C_M^G, C_m^{CA}:	averaged specific heat of flue gas resp.combustion air
$\dot{m}_G, \dot{B}, \nu\dot{B}$:	mass flow of flue gas resp. fuel/oil resp. combustion air
H_u:	caloric value of fuel/oil

The coefficients k, k_w depend on α, α_s etc. and are get by solving one dimensional nonlinear equations. A possible refinement of (1) is get by splitting $\theta(x)$ into $\theta_t(x)$ and $\theta_b(x)$.

(2) Additionally because of $\dot{m}_G = 0$ at the discharging end we have:

$$(1+\nu)\ \dot{B}.C_m^G(\theta)\theta \ = (*) \quad \text{(right hand side)}$$

(3) A mass balance for flue gas gives

$$-\frac{d}{dx}(\dot{m}_G(x)) = (1+\nu)\ \dot{B}(x)$$

(4) The recuperator (see Figure 5)

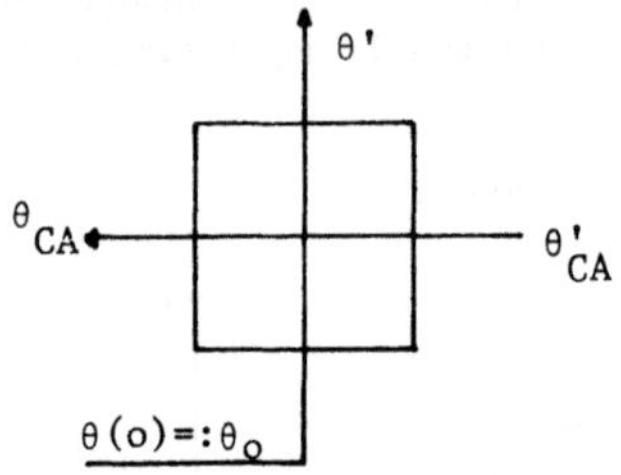

θ'_{CA}: temperature of fresh air (given)
θ': temperature of waste gas (unknown)

a) $$\dot{m}_{CA} = \frac{\nu}{1+\nu}\dot{m}_G \Big|_{x=0}$$

b) $$[C_m^G(\theta_o)\theta_o - C_m^G(\theta')\theta'] - \frac{\nu}{1+\nu}[C_m^{CA}(\theta_{CA})\theta_{CA} - C_m^{CA}(\theta'_{CA})\theta'_{CA}] = 0$$

c) $$\frac{\nu}{1+\nu}\dot{m}_G\ [C_m^{CA}(\theta_{CA})\theta_{CA} - C_m^{CA}(\theta'_{CA})\theta'_{CA}] - A_{Reku}k_{Reku}\,\Delta\theta = 0$$

$$\Delta\theta := [(\theta_o - \theta_{CA}) - (\theta' - \theta'_{CA})] \,/\, [\ln(\theta_o - \theta_{CA}) - \ln(\theta' - \theta'_{CA})]$$

(5) For the determination of T we use (M1) with $T_f := \theta(x)$

(6) objective: $\int_o^{L_x} B(x)dx \overset{!}{=} \sum_{i=1}^{4} B_i = \sum_{i=2}^{4} B_i$ (resp. $\sum_{i=2}^{4} B_i^t + \sum_{i=2}^{3} B_i^b$) = Min!

(one parameter for the flow of combustion material in zone II, III, IV)

(7) Constraints:
state variables: $T(x,y,z)$, $\theta(x)$
control variables: (B_2, B_3, B_4) <u>and</u> $(\theta_{CA}, \theta_o, \theta')$

a) We have upper and lower bounds both for the state and the control variables.
b) Equational constraints are given by 4b, 4c and the boundary condition for T.

Numerical solution: For T = T(x) the problem is to be solved from left to the right because of condition (2). For T = T(x,y) resp. T(x,y,z) this device would result in an (extremly) ill-posed problem because of the heat equation. Therefore, we decided to solve the whole problem simultaneously (at the price of a considerable numerical effort).

References:

[1] H.Engl/Hj.Wacker/E.Zarzer 'Ausgewählte Aspekte der Kontrolltheorie in computergestützten Planungssystemen' (ed.Noltemeier), Physica-verlag, pp. 155-2o2, 1976
[2] D.Auzinger/Hj.Wacker/W.Zulehner et al 'Optimierung eines Stoßofens', Report on the problemseminar "Stoßofen" (in preparation)
[3] K.L.Teo/Z.S.Wu 'Computational Methods for Optimizing Distributed Systems', Academic Press, 1984

PART IV: RESEARCH AND INDUSTRIAL CONSULTING ACTIVITIES

Permanent staff: Prof.Dr.H.W.Engl, Dr.H.Gfrerer, Prof.Dr.Hj.Wacker, Dr.W.Zulehner
Temporary staff founded by
1. Austrian Science Foundation (FWF):
a) Project S32/o3 (Engl):Dipl.-Ing.T.Langthaler, Dr.A.Neubauer
b) Project P4786, P5729 (Wacker): Dipl.-Ing.W.Bauer, Dipl.-Ing.S.Buchinger, Dipl.-Ing.E.Lindner
2. Austrian National Bank Science Foundation: Dipl.-Ing.D.Auzinger
3. OKA: Dipl.-Ing.A.Schwarz
4. VOEST-ALPINE: Dipl.-Ing.L.Peer.

Research activities: Numerical Analysis, Optimization/Opt.Control, Applied Mathematics
Special topics: Nonlinear structured highdimensional problems, inverse and ill-posed problems.
Consulting activities:
HYDRO ENERGY: A selection in addition to Lit.(problem I): (TKW, OKA)
S.Buchinger/Hj.Wacker 'Instant Optimization of Hydro Energy Storage Plants, *Operat.Res.in Progress* (ed.Feichtinger, Kall), Reidel, pp. 137-153 (1982)
S.Buchinger/H.Reisinger/Hj.Wacker 'Tagesoptimierung eines Speicherkraftwerkes auf der Basis einer Momentanoptimierung', *ZOR/B*, **25**, pp.139-157 (1981)
W.Bauer/H.Reisinger/Hj.Wacker: 'Optimale Höhensteuerung (System Schwarzach', *ZOR/B*, **26**, pp. 145-169 (1982)

W.Bauer/S.Buchinger/Hj.Wacker 'Einsatz mathematischer Methoden in der Hydroenergiegewinnung, *ZAMM* **67**, pp.227-243 (1983)
Hj.Wacker (ed). 'Applied Optimization Techniques in Energy Problems', Teubner, pp. 1-485 (1985)

STEEL INDUSTRY (VOEST) in addition to Lit. (problem II)
H.Engl/T.Langthaler 'Numerical solution of an inverse problem connected with continuous casting of steel, *ZOR/B*, Vol. **2**, (1986) (Optimal regulation of spray cooling in order to approximate a prescribed solidification front).
H.Engl/G.Landl 'Optimal scheduling in the system "continuous casting - direct rolling" (in progress)
H.Engl/A.Neubauer 'Calibration of hot-rolling mills (in progress)
W.Zulehner 'Hydrodynamic coefficients for turbines (VOEST) 1978, Dipl.Thesis
F.Bachmayr 'Singular FE for cracking problems, package ADINA (VOEST) 1982, Dipl.-Thesis
R.Seil 'Semi infinite FE (package ADINA) (VOEST) 1983, Dipl.-Thesis
R.Feichtl 'Optimal design of technical constructions (VOEST) 1984, Dipl.-Thesis
B.Lindorfer 'Optimal design of girders (VOEST) 1984, Dipl.-Thesis

Industrial cooperation with various other companies (selection)
K.Mair 'Parallel Numerical Techniques for the System SMS 2o1 (Siemens München), 198o, Dipl.-Thesis
E.Walchshofer 'Forecasting the daily load by Cluster analysis and Learning Machines (Siemens, Vienna) 1985, Dipl.-Thesis
M.Kowalik/E.Zarzer/W.Zulehner 'Zur Gestaltung eines verlustarmen Spiralgehäuses für Kreiselpumpen (Fa.Ochsner), Forsch.Ing. Wes., Nr.6
H.Gfrerer 'Geometry of wire ropes' (Fa.Teufelberger), Arbeitsbericht: Berechnung von Litzen der Normalmachart, S. 1-25, 1985
D.Auzinger/S.Buchinger 'Optimal production scheme for kitchen furniture (Fa. EWE), Studie pp. 1-48
H.Engl/E.Lindner 'A combined boundary value and transmission problem arising from the calculation of eddy currents (in reactor coils), (Spezielektra-Esslinger), *Journ.Appl.Math./Phys.* **35**, pp.289-3o7 (1984)
E.Lindner/H.Petritsch/Hj.Wacker/E.Walchshofer 'Lastprognose' (OKA), Studie pp. 1 - 24.
R.Takacs 'Ein Simulationsmodell für den Straßenverkehr'(Stadt Krems), doct.thesis, Institut für Mathematik, Universität Linz, 198o

Acknowledgements: Research concerning problem I and II was partially supported by the Austrian Science Foundation FWF (P4786, P5729). Research concerning Problem III is partially supported both by the FWF (P5729) and the Austrian National Bank Science Foundation (Nr.258o).

A Collection of Capsule Descriptions of Mathematics-in-Industry Problems

Below there are some 50 pages of short descriptions of various problems from industry which have been tackled mathematically. The collection is intended to give a first weak indication of just how many such applications there are and how varied they are. There is of course much much more and neither the quantity nor the variety are adequately suggested by these pages. As a matter of fact more material was provided by the participants to ESMI. For instance a number of five line descriptions by the group of Prof. Y. Cherruault, Medimat, Paris, dealing with optimal control of fermentation processes, clarification of apple juice and (very important?) cider, fraud detection in juices and wines, automatic treatment of ultrasonic signals (imaging), workshop management in the food industry, justification of homeopathic drugs, effect of vibrations on the spine and the optimization of diabetic treatments.

Eventually ECMI hopes to undertake the task of compiling an adequate well cross-referenced catalogue of industrial applications of mathematics. Such a thing is definitely needed. Even at this modest conference it became clear that many times the same problems are solved over and over again at different places and (more interestingly) more or less the same problems are solved differently.

MULTISTAGE MULTICOMPONENT CHEMICAL COMLUMNS

Problem. The separation of a mixture of liquid and volatile components may be performed f.e. by multistage distillation or absorption techniques. Extended models also allow main columns with sidestrippers and/or pumping activities. The model describes the thermodynamic equilibrium of the mixture. Further extensions concerning more complex chemical plants can be handled by splitting techniques, resulting from cutting flows between chosen subsystems.

Mathematics. Using the component material balances, the enthalpy balances and the conditions for the liquid-vapour equilibrium one gets a highly nonlinear large scaled system with special structure. In the simplest case of one separation column one gets a block three diagonal Jacobian. Using a special hybrid technique including damping, projection, and imbedding, the problem can be solved quite efficiently either by Newton or Quasi-Newton techniques. The extended cases of sidestrippers and/or pumping activities give rise to off-diagonal blocks. They are treated either by Newton or by Newton-SOR.

[1] D. AUZINGER, Computation of the thermodynamic equilibrium in distillation/absorbtion columns by Quasi Newton techniques, Diploma Thesis, Dept. of Mathematics, University Linz, 1985.

[2] L. PEER, Computation of the thermodynamic equilibrium in distillation/absorption columns by Newton techniques, Diploma Thesis, Dept. of Mathematics, University Linz, 1985.

[3] D. AUZINGER, F. KOKERT, L. PEER, Hj. WACKER, W. ZULEHNER, Multistage Multicomponent Chemical Columns, in preparation (1986).

For more information contact D. Auzinger, L. Peer, Hj. Wacker, W. Zulehner, Working group on industrial and business mathematics, Universität Linz, Austria; F. Kokert, Voest-Alpine AG, Linz, Austria.

OPTIMAL REHEATING OF SLABS IN PUSHER TYPE FURNACES

Problem. In steel processing reheating of slabs for rolling is very energy consuming. Thus one wants to control the reheating process in such a way that all technical requirements are fulfilled and energy costs become a minimum. In a first step we confine ourselves to a steady state model with constant charging temperature and charging speed of the (uniform) slabs.

Mathematics. The steady state model consist of the heat equation for the slabs (in one, two or three dimensions, nonlinear), a bilance model for the flue gas and the combustion processes and some nonlinear equations modelling the recuperator where the combustion air is preheated by the waste gas. By controlling the fuel input energy costs can be minimized. The resulting numerical model is a highly nonlinear, constrained optimization problem. Since for each evaluation of the constraints the parabolic system for the temperature distribution of the slabs and the bilance equations for the flue gas have to be solved, which is very time consuming, a very efficient optimization algorithm must be used. The parabolic system with one spatial variable is solved by the Crank-Nicolson method. To discuss the influence of the water-cooled skids a system with two spatial variables is studied.

[1] D. AUZINGER, B. LINDORFER, Hj. WACKER, W. ZULEHNER, Optimal Reheating of Slabs in a Pusher Type Furnace, ZAMM (to appear), T1 - T4.

For more information contact D. Auzinger, Hj. Wacker, W. Zulehner, Working Group on Industrial and Business Mathematics, Universität Linz, Austria; B. Lindorfer, Voest-Alpine AG, Linz, Austria.

OPTIMAL PRODUCTION SCHEME FOR THE GOSAU STORAGE POWER PLANT SYSTEM

Problem. The System Gosau consists of three serially connected storage power plants each with a reservoir of different storage capacity. The aim was to maximize the annual production of energy with respect to a number of constraints both technical and other ones. The largest reservoir, Lake Gosau, is not watertight because it is situated in a chalk formation of the Alps. There are heavy seepage losses depending on the actual height of the reservoir. During summer the level of the Lake is to be held above a certain height by demands of the tourist trade. A supply treaty with the Austrian Railway Company requires that 1.5 [MW] are of be produced as priority at the third power plant from 6 a.m. to 10 p.m.

Mathematics. Modelling was performed in cooperation with engineers of the Upper Austrian Power Board Company. One gets a nonlinear nonconvex problem of optimal control. Though it was possible to split up the annual optimization into three independent parts there result large scaled problems ($\sim$ 700 variables, 2800 constraints). Besides classical dynamic programming [1] we used a newly developed decomposition/convexification procedure [2] for the numerical solution. By means of variational techniques the dimensionality of the problem could be reduced further. Sensitivity analysis was performed by simulation.

Application. Our results have been applied in practice during the last three years. Comparing these years with similar years (depending on the annual influx) we got an increase of energy production of about 15%.

[1] Hj. WACKER: Applied Optimization Techniques in Energy Problems, Proceedings, Workshop, Linz, June 1984, Teubnerverlag, 1985, pp. 1-485.

[2] H. GFRERER: Globally Convergent Decomposition Methods for Nonconvex Optimization Problems, Computing 32, (1984) pp. 199-227.

For more information contact W. Bauer, H. Gfrerer, E. Lindner, A. Schwarz, Hj. Wacker, Working Group on Industrial and Business Mathematics, Universität Linz, Austria.

OPTIMIZATION OF HYDRO ENERGY STORAGE PLANT SYSTEM BY VARIATIONAL METHODS

1. Problems. In the general framework of hydro energy optimization we consider the question of maximizing of the value of production of a storage plant system within a given interval of time. Assuming different tariff periods one gets a nonlinear convex optimization problem with constraints. The simplified case of constant efficiencies (generalization: efficiencies not depending on the discharges) is solved semi-analytically for two special problems:

i) optimal production for a sequential system of storage plants [2]

ii) optimal production of an one storage plant system in the presence of spilling water [3]

2. Mathematics. In both cases the problem is discussed as a problem of optimal control, where the guiding equations are given by the continuity equations. The box constraints for the state variables (the volumes of the reservoirs) are included by Lagrange Multipliers, the box constraints for the controls (the discharges) are respected directly when optimizing. Using the fact that the objective depends only linearly on the discharges one succeeds to transform the infinite dimensional problem into an equivalent finite dimensional one which may be solved classically or by [4].

[1] H. GFRERER, Optimization of Hydro Energy Storage Plant Problems by Variational Methods, ZOR/B 28, pp. 87-101 (1984).

[2] W. BAUER, Optimal Control of Storage Power Plants, Doctorial Thesis, 1986.

[3] E. LINDNER, Modelling and Optimal Control for Storage Plants, Doctorial Thesis, 1986.

[4] H. GFRERER, J. GUDDAT, Hj. WACKER, A globally convergent algorithm based on imbedding and parametric optimization, Computing 30, pp. 225-252 (1983).

For more information contact W. Bauer, H. Gfrerer, E. Lindner, Hj. Wacker, Working Group on Industrial and Business Mathematics, Universität Linz, Austria.

OPTIMAL DAILY PRODUCTION SCHEME FOR STORAGE POWER PLANTS

1. Problem. The power distribution for a full day's production of an large hydro energy power plant is approximately known in advance, say by a prognosis. Given the prognosis, an optimal production scheme for the whole day is proposed resulting in a minimum loss of the total potential energy.
For a n-machine power plant one has to decide the mode and the activity level of each machine during each time interval $T_1, \ldots, T_N$. A machine cen be at rest, running at zero level (no starting costs!), or producing at nonzero level. There are losses 1) in the pipe system, 2) in the turbines, 3) in the generators, 4) by switching, 5) by running a machine at zero level.

2. Mathematics. First the problem of instant optimization is solved off-line [1]. A given load P is distributed to all active machines and the losses of potential energy are determined. This has to be done for all possible modes of the n-machines set for all possible heads. Mathematically this leads to a n dimensional constrained optimization problem which may be solved by a special homotopy technique [2] using P as parameter for the imbedding. Using the results of the first model the N-stage optimization problem determining the mode and the activity level of each machine is solved on-line by a special dynamic programming technique. For a 4-machine power plant one needs less than ten (IBM 370/155) for one day [3].

[1] S. BUCHINGER, Hj. WACKER, Instant Optimization for Hydro Energy Generation in Storage Plants, Proc. ÖGOR Vienna 1980, Reidel 1982, 137-145

[2] H. GFRERER, J. GUDDAT, Hj. WACKER, A Globally Convergent Algorithm Based on Imbedding and Parametric Optimization, Computing 30, 225 - 252 (1983)

[3] S. BUCHINGER, H. REISINGER, Hj. WACKER, Tagesoptimierung eines Speicherkraftwerkes auf der Basis einer Momentanoptimierung, ZOR/B, Bd. 25 (1981) B139-B157

For more information contact S. Buchinger, H. Reisinger, Hj. Wacker, Working Group on Industrial and Business Mathematics, Universität Linz, Austria

BOOLEAN APPROXIMATION IN PRODUCT SPACES

Methods of multivariate interpolation and approximation are important for many fields of applied mathematics such as computer aided geometric design (car production), finite elements, and multivariate data fitting in geophysics.
A number of papers appearing in the years since 1969 have been concerned with developing the algebraic aspects of "blending approximation" (Gordon [1], Gordon-Cheney [2], Delvos [4], Delvos-Posdorf [3]). This theory has been proved useful in the task of synthesizing multivariate interpolation operators from univariate for various special purposes. In the recent papers (Delvos-Schempp [5,6]) a systematic treatment of the blending method was presented in the framework of functional analytic approximation theory.
The Boolean method of approximation is used to construct polynomial interpolation schemes in many variables which are used in the finite element method (Gordon-Hall [8], Delvos-Posdorf [7]). Boolean interpolation enables the generation of multivariate polynomial Hermite interpolation schemes which are important in the finite element method and in the field of computer aided geometric design. Recently we applied Boolean methods to construct global schemes of bivariate Hermite interpolation for smooth periodic functions in two variables using trigonometric polynomials in two variables (Delvos [9]). For this method the approximation power is studied.

[1] W.J. GORDON, Distributive lattices and the approximation of multivariate functions. In: Approximation with special emphasis on spline functions (I.J. Schoenberg, ed.), Academic Press, New York, 1969.

[2] W.J. GORDON, W.W. Cheney, Bivariate and multivariate interpolation with noncommutative projectors. In: Linear spaces and approximation (P.L. Butzer and B.Sz. Nagy, eds.), ISNM 40 (1977), Birkhäuser Verlag, Basel, 1977.

[3] F.-J. DELVOS, H. POSDORF, N-th order blending. In: Constructive theory of functions of several variables (W. Schempp and K. Zeller, eds.), Lecture Notes in Mathematics 571 (1977).

[4] F.-J. DELVOS, d-variate Boolean interpolation. J. Approximation Theory 34 (1982), 99-114.

[5] F.-J. DELVOS, W. SCHEMPP, On precision sets of interpolation projectors. In: Multivariate approximation theory II (W. Schempp and K. Zeller, eds.), ISNM 61 (1981), Birkhäuser Verlag, Basel, 1982.

[6] F.-J. DELVOS, W. SCHEMPP, The method of parametrix extension applied to right invertible operators. Numer. Funct. Anal. and Optimiz. 6 (1983), 135-148.

[7] F.-J. DELVOS, H. POSDORF, Boolesche zweidimensionale Lagrange-Interpolation. Computing 22 (1979), 311-323.

[8] W.J. GORDON, C.A. HALL, Transfinite element methods: Blending function methods over arbitrary curved element domains. Numer. Math. 21 (1973), 109-129.

[9] F.-J. DELVOS, On bivariate Hermite trigonometric interpolation. In: Constructive theory of functions '84, Sofia 1984, p. 226-272.

For more information contact Prof.dr. F.-J. Delvos, Mathematik I, Universität Siegen, D-5900 Siegen, FRG.

MULTIVARIATE AUTOREGRESSIVE MODELING - A NEW TOOL FOR TROUBLE-SHOOTING MEASUREMENTS

Clearing up and elimination of disturbances in process of production is daily work of operating and maintenance personnel. This work is based on information provided by the normal instrumentation of the plant and on the practical experience of the personnel. Special measurements and signal processing methods, which are needed when this normal work can't give sufficient results, are called trouble-shooting diagnostics.

The quality and uniformity of the product can be used as measure of undisturbedness and goodness of the process. In addition to that it is important to minimize thermal and mechanical stresses to the process components. Harmful disturbances can be caused e.g. by vibration problems or by inadequate operation of control loops.

It is very difficult to use conventional signal analysis methods - one or two channel analysis - in trouble shooting diagnostics. We can't get any reliable information about the interaction of the signals, which describes the internal dynamic structure of the process. Multivariate autoregressive modeling can give us that information. It is dedicated to solve such problems as noise source identification or evaluation the cause-effect relationships between measured signals.

A multivariate autoregressive model (MAR) defined in matrix form is

$$x(k) = \sum_{m=1}^{M} a(m)x(k-m)+e(k)$$

$$x = [x_1,x_2,x_3,\ldots,x_n]^T$$

$$a = \begin{bmatrix} a_{11} & a_{12} & \cdots & a_{1n} \\ A_{21} & a_{22} & \cdots & a_{2n} \\ \vdots & \vdots & & \vdots \\ a_{n1} & a_{n2} & \cdots & a_{nn} \end{bmatrix}$$

$$e = [e_1,e_2,\ldots,e_n]^T$$

e consists of white noise processes, M is the model order and n is number of measured signals.

The MAR-model represents a system with all variables explaining each other. The values of each variable are explained by former values of all variables and white noise processes. This kind of approach is applicable to the analysis of systems with feedback loops.

The MAR-model can be estimated using covariance functions which have been calculated from the measured signals. After estimating the discrete parameter signal model we can derive such results as the auto spectral density functions, of the signals and the noise contributions, which describe the contribution of each noise source to the spectrum of the examined signal.

The Central Laboratory of Imatran Voima Oy has developed in co-operation with Tampere University of Technology a program package, which applies MAR-method in use. This highly interactive program package has been written in FORTRAN-77 for VAX-11 computers.

For more information contact Jari Eklund, Imatran Voima Oy, IVO Central Laboratory, P.O. Box 112, 01601 Vantaa, Finland.

FORECASTING PENSION PAYMENTS BY AUSTRIAN BANKS FOR THE NEXT 25 YEARS

Problem. Austrian banks pay their employees pensions after retirement that usually exceed the public social security pensions. Both the age structure of the employees and cuts in the public social security system lead to the expectation that the strain on the banks' pension funds will increase dramatically in the next 25 years. The program developed allows a (probably rather accurate) quantitative study of this development. It also allows to study the effects of various changes in the agreements that regulate the payment of bank pensions. The program has been implemented for a group of 11 Austrian banks.

Mathematics. Simulation

Exact results are contained in confidential reports. Short survey: Risk Kontroll 1 (1985), Vienna.

For more information contact Heinz W. Engl, Working Group on Industrial and Business Mathematics, Universität Linz, Austria.

CONTROL OF SOLIDIFICATION IN CONTINUOUS CASTING OF STEEL

Problem. Continuous casting is the dominant process for solidifying liquid steel. In this process liquid steel is cast into the mould, which must extract sufficient heat from the incoming metal to solidify a shell which can support the liquid pool at the mould exit. The heat extraction process is continued by spraying water onto the surface of the strand to remove sufficient heat from the steel to virtually complete solidification. It is of major importance for the quality of the resulting product how the front of solidification behaves as a function of time. Hence one is interested to control the process (via the cooling rates determined by the water pressure) in such a way that the resulting solidification front is as close as possible to an optimal one.
We call the problem of computing the resulting front of solidification, given the cooling rates, the direct problem. The problem of computing cooling rates such that the resulting front of solidification approximates an optimal one as closely as possible will be denoted by inverse problem.

Mathematics. The mathematical formulation of the continuous casting process leads to a nonlinear heat equation with nonlinear mixed (Neumann and Dirichlet) boundary conditions. The direct problem is similar to a free boundary problem of Stefan type. Under certain, sometimes restrictive, though quite reasonable, assumptions one can process existence, uniqueness and continuous-dependence results for the direct problem. The inverse problem is a general Inverse Stefan Problem which is even for a linear heat equation ill-posed in the sense, that the solution (i.e., the cooling rates) does not depend continuously on the data (i.e., the solidification front). Further assumptions make it possible to prove uniqueness and restore continuous dependence for the inverse problem. For mathematical details see [1].
In order to solve the inverse problem numerically, one has to regularize this ill-posed problem. In the concrete technological environment for which we developed our method, this regularization was built into the model for technological reasons: The cooling rates in the casting machine we treated could only be controlled in six different zones; within each of these zones, the cooling rate had to be constant. Thus, the cooling rates could be treated as a six-dimensional variable and hence, the inverse problem was reduced to a finite-dimensional problem of nonlinear approximation. For the algorithm and a report about the implementation in an industrial environment see [2].

[1] H.W. ENGL, T. LANGTHALER, P. MANSELLI, *On an Inverse Problem for a Nonlinear Heat Equation Connected with Continuous Casting of Steel*, to appear in K.H. Hoffmann - W. Krabs (eds.), Optimal Control with Partial Differential Equations, Birkhäuser, Basel 1987.

[2] H.W. ENGL, T. LANGTHALER, *Numerical Solution of an Inverse Problem Connected with Continuous Casting of Steel*, ZOR-Zeitschrift für Operations Research, Volume 29 (1985) pp. B185 - B199

For more information contact Heinz W. Engl and T. Langthaler, Working Group on Industrial and Business Mathematics, Universität Linz, Austria.

OPTIMIZATION OF THE BALANCE-SHEET STRUCTURE OF A BANK

Problem. Starting from one year's average balance sheet, the program determines the structure of next year's average balance sheet that yields the optimal profit under all relevant restrictions (e.g.: legal restriction, upper and lower bounds for growth of single positions or combinations of positions, bounds for relations of certain groups of positions).
Another program can be used for optimizing the balance sheet structure for a three-year horizon; relevant data are forecast from assumptions about the development of general economical quantities. The programs have been implemented at 2 Austrian banks.

Mathematics. Linear programming, regression

The results will be submitted to Zeitschrift für Operations Research, Serie B, in late 1986.

For more information contact H.W. Engl, C. Lindl, B. Weigerstorfer, Working Group on Industrial and Business Mathematics, Universität Linz, Austria.

CALCULATION OF EDDY CURRENTS IN REACTOR COILS

Problem. A company based in Linz that manufactures large reactor coils (for reducing start-up currents, limiting short-circuit currents in rectifier equipment, limiting the current under fault conditions etc.) needed a computer program for calculating losses caused by eddy currents; this program should run on their relatively small computer. The software developed achieves this goal.

Mathematics. The problem leads to a combined boundary value and transmission problem for a combination of Laplace's and Helmholtz' equation. This problem was treated by a combination of a finite-difference method and a boundary integral equation method.

Reference. A combined boundary value and transmission problem arising from the calculation of eddy currents: Well-posedness and numerical treatment, Zeitschrift f. Angew. Math. und Physik 35 (1984), 289-307.

For more information contact H.W. Engl, E. Lindner, Working Group on Industrial and Business Mathematics, Universität Linz, Austria.

HEATING OF INTUMESCENT PAINTS: AN IRREVERSIBLE FREE BOUNDARY PROBLEM

Heating of an intumescent paint results in an outgassing process accompanied by considerable swelling and absorption of latent heat occurring over a narrow temperature interval. Intumescent paints are used as a protective coating against external heat sources. The swelling region can be modelled as a front. As long as the heat input is sufficiently intense to sustain the process, the corresponding mathematical model is nothing but a two-phase Stefan problem (see [1] and the literature quoted therein). However the process obviously involves irreversible change of phase processes. An effort in such a direction can be found in [2] (see also e.g. [3] for change of phase with hysteresis) and in [4].

[1] J. BUCKMASTER, C. ANDERSON Jr., A. NACHMAN. The response of intumescent paints to heat. In *Free Boundary Problems: Theory and Applications*, **3**, **4** (Bossavit, Damlamian, Frémond eds.), Res. Notes Math., Pitman.

[2] E.DI BENEDETTO, A. FASANO, M. PRIMICERIO. An irrevesible free boundary problem. To appear on Control and Cybernetics.

[3] A. VISINTIN. A phase transition problem with delay. Control and Cybernetics **11** (1982), 5-18.

[4] M. FREMOND, A. VISINTIN. Dissipation dans le changement de phase. Surfusion. Changement de phase irréversible. To appear on C.R. Acad. Sci. Paris.

For more information contact A. Fasano, Mathematical Institute, University of Florence, Italy.

A MATHEMATICAL MODEL FOR SOLIDIFICATION PROCESSES

It is well known that e.g. in continuous casting processes a significant supercooling is observed, accompanied by the formation of dendrites in the vicinity of the solidification front. Interesting models have been proposed either describing the growth of dendrites [1] or introducing a thickness of the solid-liquid interface on the basis of thermodynamical arguments [2]. The first attempt is micro-scale oriented, while the second one is more in the spirit of an averaging process, avoiding sharp boundaries.

The model proposed below is somehow in between, since it takes into account the average effect of surface tension in the dendritic zone and it gives an implicit description of the boundaries of completely liquid and completely solid regions. Referring to a pure substance, we write the following equations

$$k\Delta T - c\dot{T} = -L\dot{s}, \tag{1}$$

$$h\dot{s}^{+} = (T/T_0 - f(s))H(T/T_0)H(1-s), \tag{2}$$

$T =$ temperature ($T=0$ is the melting temperature in normal conditions), $s =$ solid fraction, $f(s) \sim s^m (m>0), T_0<0$ is super-cooling under stationary conditions, H is the Heavyside function, k, c, L are obviously defined, h is some positive constant, the dot denotes Lagrangian derivative.

The term $f(s)$ in equation (2) expresses the influence of surface tension due to the average curvature of the liquid-solid interface in the dendritic zone at equilibrium. The term $h\dot{s}$ (positive part) brings into the model the influence of the solidification rate on the average curvature.

A properly defined weak solution can be shown to exist for (1), (2) possessing some expected qualitative properties [3].

Numerical computations are being performed in order to test the model by comparison with experimental data for continuous casting processes.

[1] J. OCKENDON, A.B. TAYLER. A model for alloy solidification. In *Free Boundary Problems: Applications and Theory*, A. Bossavit, A. Damlamian, M. Frémond eds. Res. Notes Math. (Pitman) **120**, 157-165.

[2] G. CAGINALP. Phase field models of solidification: free boundary problems as systems of nonlinear parabolic differential equations. Ibid. 107-121.

[3] A. FASANO, M. PRIMICERIO. In preparation.

For more information contact A. Fasano, Mathematical Institute, University of Florence, Italy.

GEOMETRY OF WIRE ROPES

Problem. In general a wire rope consists of a number of strands wound around a center core where again each strand consists of several wires wound around a center wire in several layers. For the design of wire ropes with high service lift geometrical aspects, e.g. the interference and clearance of adjacent wires, are of great importance. In cooperation with an Upper Austrian wire rope manufactory a computer program was made for calculation of the required geometrical quantities.

Mathematics. By use of differential geometry the problem may be formulated as a system of non-linear equations which is solved by standard techniques.

Theoretical foundations and programs are confidential.

For more information contact H. Gfrerer, Working Group on Industrial and Business Mathematics, Universität Linz, Austria.

ON THE CONSTRUCTION OF THE VOLUTE OF A CENTRIFUGAL PUMP

Problem. For constructing the volute of a centrifugal pump either the principle of conversation of the angular momentum according to C. Pfleiderer or the principle of conservation of kinetic energy according to A.J. Stepanoff are often used. In this work both principles are combined and additionally the hydraulic diameter of the cross sections of the volute are maximized for minimizing friction losses.

Mathematics. This leads to a classical problem in the calculus of variation with two constraints and a variable end point for the cross sections of the volute. The Euler-Lagrange equation and the transversality conditions for this problem are reduced to a system of three nonlinear equations in three unknowns, which was numerically solved by Newton's method.

[1] M. KOWALIK, E. ZARZER, W. ZULEHNER, Zur Gestaltung eines verlustarmen Spiralgehäuses für Kreiselpumpen, Forsch. Ing.-Wes. 46 (1980), pp. 196-198.

For more information contact M. Kowalik, E. Zarzer, W. Zulehner, Working Group on Industrial and Business Mathematics, Universität Linz, Austria.

FORECASTING OF THE DAILY LOAD

The progress line of the daily load is of great importance to all suppliers of electric energy as it influences the operation mode of each power plant. A prediction of the daily load (normally 48 values, representing 48 time intervals of 30 minutes) is used by loads dispatchers as a decisive factor for cutting, for keeping in standby mode or for restarting certain turbines or power plants. It is also used as an input parameter for daily optimization problems of power plants.

At first characteristic progress lines for the daily load are determined by cluster analysis and conditional probabilities for the occurrance of these characteristic progress lines under certain weather conditions are computed by using learning machines. Based on these an actual forecasting is calculated. The method has been applied to two power plant companies.

Literature:

[1] E. WALCHSHOFER: Forecasting of the Load by Using Cluster Analysis and Learning Machines (in German), diploma thesis, Univ. Linz, 1985, 1-93.

For more information contact E. Lindner, Hj. Wacker, E. Walchshofer, Siemens Austria AG, Vienna: G. Petritsch, Working Group on Industrial and Business Mathematics, Universität Linz, Austria.

INPUT PARAMETERS FOR OPTIMIZING HYDROELECTRIC POWER PLANTS

Our main goal for hydroelectric storage power plants is to maximize the value of the energy produced within a certain interval of time, say one day or one week. For this we have to solve a nonlinear non-convex optimization problem. To put up this model, however, one has to determine some input data. A few of those give rise to some trouble.

The first one is the efficiency function of the power station. The main idea is to model the physical situation, i.e., to consider the losses in the tunnel system, in the turbines and in the generators and use physically motivated polynomial approximations. This leads to a nonlinear problem of parameter identification. Additionally not only the measurements but also the points where the measurements are taken are subject to errors. For overcoming this difficulty we used an idea of Schwetlick/Tiller [4]. This method is based on a generalized least squares technique and takes advantage of the certain structure of the Jacobian in the optimization process. The algorithm has successfully been applied to the power plants Partensein and Gosau (see [1]).

A second difficulty arises from the influx. In case of heavy rainfall there is considerable loss of water running over the weir in Partensein. In [1] a multivariate normal distribution was used for a first model. Recently (cf. [2], [3]), also a transfer function model of the type

$$A(q^{-1})\tilde{z}(t) = B(q^{-1})U(t) + G(q^{-1})R(t) \text{ and } Z(t) = \tilde{z}(t) + \xi(t)$$

has been applied for predicting the influx. Here, $R(t)$ are gauge data, $U(t)$ are transforms of rainfall and temperature, A, B and G are polynomials in the timebackshift operator q^{-1}, and $\xi(t)$ is a noise of type $ARMA(1,1)$. The parameters of the model were estimated by a cursive instrumental variable method from the data of one month for Parenstein whereas the predictions were made for another 2, 4 and 6 hours ahead by the Kalman filter method.

[1] E. LINDNER, Hj. WACKER, Input Parameters for the Optimization of the System Partenstein: Forecasting of the Influx, Determination of the Efficiency Function, in: Hj. Wacker (Ed.): Proceedings of the Conference Applied Optimization Techniques in Energy Problems, B.G. Teubner Stuttgart 1985, 286-320.

[2] E. LINDER, G. OBERAIGNER, Hj. WACKER, A. WAKOLBINGER, The Hydroelectric Power Plant Partenstein: Forecasting Determination, to appear in Proceedings of "18th Annual Conference on Mathematical Optimization", Diedrichshagen, GDR, 1986.

[3] G. OBERAIGNER, Influx Prediction for the Reservoir of the Power Plant Partenstein (in German), diploma thesis, Univ. Linz, 1986, 1-144.

[4] H. SCHWETLICK, V. TILLER, Numerical Methods for Estimating Parameters in Nonlinear Models with Errors in the Variables 27(1), 17-24, 1985.

For more information contact E. Lindner, G. Oberaigner, Hj. Wacker, A. Wakolbinger, Institut für Mathematik, Universität Linz, Austria.

THE PROBLEM OF THE NON-STATIONARY BALLOON IN TEXTILE PROCESSES

In some textile processes (e.g. winding) a yarn passes through an eye-guide and is dragged by some rotating device. As a consequence the free portion of the yarn assumes a particular configuration at each time t (a skew curve) and described a surface, which is known as the balloon related to the process. Determining the shape of the balloon is of practical importance. The balloon can be non-stationary if the conditions at the spinning device are time dependent (as in the case of winding). The problem of calculating the balloon can be modelled in a rigorous way as a system of nonlinear hyperbolic equations with a free boundary. The presence of a free boundary is motivated by the fact that the length of the curve generating the balloon is a priori unknown, as well as some elements of the motion of the ring guiding the yarn. Boundary data are also very complicated: for instance the angle between the yarn and the reel satisfies a Riccati equation [11-12].

Owing to the complexity of the problem only some numerical treatment is available. The mathematical investigation of the influence of the various parameters entering the problem on the final product is a completely open question.

[1] PADFIELD D.G., A Note on the Fluctuations of Tension During Unwindings, Journal of Textile Institute, Vol. 47, N.6, 1956 pp. T301-T303.

[2] TURTELTAUB M.J. and M.A. BEJAR, Balloon Collapse in Ring-Spinning, ASME Journ. of Eng. for Industry, Vol. 98, N.2, 1976 pp. 658-663.

[3] TURTELTAUB M.J., et al., Calculation of Peak Yarn Tension During Ring-Spinning, ASME Journ. of Eng. for Industry, Vol. 100, N.2., 1978 pp. 167-170.

[4,5,6,7,8] KOTHARI V.K., et al., Journal of Textile Institute, Vol. 70, N.3, 1979, pp. 89-95 — Vol. 70, N.3, 1979, pp. 96-104 — Vol. 70, N.5, 1979, pp. 172-183, pp. 193-197.

[10] CITTI P., et al., Un metodo per la simulazione della filatura ad anello, Atti del VI Congresso Nazionale AIMETA-Genova, 1982, Sez.III, Vol. I, pp. 299-309.

[9] LISINI GG., et al., Sul problema non stazionario del balloon nelle machine tessili, Atti dell 'Università di Pisa, 1983.

[12] GIUSTI R., et al., Sulla determinazione del balloon nella filatura a ring, Atti del VII Convegno AIMETA, 1984, Sez. III, pp. 472-484.

For more information please contact prof. G. Lisini, prof. D.Quilghini or prof. P. Toni, Instituto di Matematica Applicata G. Sansone, Facoltà di Ingegneria, via di S. Marta 3, 50139 Firenze (Italy).

COMBUSTION OF POLYMERS WITH FLAME RETARDANTS

Polymers treated with suitable impurities develop a solid porous layer after ignition, which grows, retarding and eventually stopping combustion. The thermal regime is extremely complex, since it is influenced by several coupled phenomena, such as pyrols, vaporization and the chemical reactions occurring simultaneously in combustion and in the production of the thermally insulating porous crust. See e.g. [1] for details. A mathematical model is missing, but a first approach is in progress [2].

[1] A. BALLISTRERI et al., Intumescent flame retardants for polymers, I. J. Appl. Polymer Sci. 28 (1983) pp. 1743-1750.

[2] A. FASANO, P. FIFE, R. RICCI, In progress.

For more information contact prof. G. Montaudo, Instituto Dipartimentale di Chemica e Chimica Industriale, Viale A. Doria 6, 95125 Catania (Italy).

MODELLING AND CONTROL OF SPATIAL POWER OSCILLATIONS IN NUCLEAR REACTORS

large nuclear reactor systems can exhibit underdamped stable or unstable oscillations in the power density distribution due to physical positive feedback mechanisms within the plant. It is very important to device control strategies to damp out these oscillations in a form compatible with overall plant structure, available control mechanisms and constraints. The mathematical content of such studies is very large as safety considerations dictate that they must be essentially model based. A detailed model typically consists of several partial differential equations in 3 space dimensions and time but, for control studies, this must be replaced by a finite dimensional model of order (typically) $n=240$ using a mixture of modal, nodal and finite difference approximations. Such an approximate model has some numerical complexity due to its dimension and singular differential structure although some of the difficulties can be offset by use of the natural sparseness of the matrices. Control studies can proceed by computation of the systems transfer function matrix using extended QR and QZ methods augmented by capitalization in the natural symmetric model symmetric model structure of the plant. The final control design can then be based on feedback stabilization of practical, easily implemented form using the group symmetry properties of the system and control theoretical methods.

For more information contact Prof. D.H. Owens, University of Strathclyde, Department of Mathematics, Livingstone Tower, Richmond Street, Glasgow Gl 1XH, tel. (041)-55-4400 Ext(3804).

COMPOSITION MODELLING AND CONTROL FOR BINARY DISTILLATION COLUMNS

Multi-loop binary distillation columns are represented by a large number of nonlinear differential equations of some complexity. Despite this fact the observed in- and output performance between flow actuators and composition measurements has a deceptively simple form. Using this fact, it is natural to conjecture that the plant can be modelled, for the purpose of control systems design, by an approximate model of a very simple form. To formalize this intuition, it is necessary to construct a rigorous approximation theory and model reduction procedure suitable for control systems design in such circumstances. Such a theory should predict conditions under which a controller designed for the approximate model will also stabilize the real plant. It should also produce predictions of the likely performance deterioration due to the modelling errors. Such an approach has been developed using relatively sophisticated methods from functional analysis, fixed point theory and control theory and, despite the sophistication, is highly suitable for computer-aided-design. Practical tests using step response data from an actual column indicate that highly successful designs are possible with considerable savings in time.

For more information contact Prof. D.H. Owens, University of Strathclyde, Department of Mathematics, Livingstone Tower, Richmond Street, Glasgow G1XH, Tel.: (041) 522-4400 Ext(3804).

MODELLING AND CONTROL OF ENVIRONMENT SYSTEMS

The control of temperature and humidity is a common problem in agricultural systems and computer environments and is attempted by a mixture of moisturizing and colling mechanisms together with inlet and output dampers to the outisde air. These mechanisms are supervised by electronic control elements of simple proportional plus integral form. The choice of proportional bands and reset times for the consequent multi-loop control scheme is a complex problem that, for maximum energy efficiency, should be based on a process model of some complexity. Typical models are of nonlinear form and contain equations describing the temperature variations in major environmental spaces, delays in transit through conduit, nonlinear damper characteristics and moisturizing mechanisms together with associated actuators such as constant velocity motors and the nonlinear measurement characteristics of temperature and humidity transducers. Despite these complexities, a systems theoretical approach to the modelling problem can lead to models of sufficient accuracy to make off-line control design and performance assessment both successful and rapid and have the bonus of enabling the comparison of different control philosophies at little extra cost.

For more information contact Prof. D.H. Owens, University of Strathclyde, Department of Mathematics, Lingstone Tower, Richmond Street, Glasgow G1XH, Tel.: (041) 552-4400 Ext(3804).

ELECTROTHERMAL ATOMIZATION IN SPECTROSCOPY

Atomic spectroscopy techniques are extensively employed in analytical chemistry. A typical electrothermal atomization procedure involves the rapid heating of a solid on whose surface the analyte is initially placed. The time-dependent concentration of the analyte in the inert gas surrounding the solid is then monitored by measuring the absorption of an optical beam. A proper mathematical model for this process is expected to suggest improvements of the operational procedures and of the equipment, as well as to allow a better understanding of the process. The model must account for the interplay of four mechanisms: release of the analyte from the surface of the solid rod; redeposition; convection in the gas; diffusion in the gas. Typically one obtains a parabolic (linear) convection-diffusion equation coupled to a (non-linear) ordinary differential equation. Singular perturbation and numerical approaches for the solution of the problems have been presented in [1-5]. A Monte-Carlo treatment has also been sugested in [6].

[1] S.L. PAVERI-FONTANA, G. TESSARI and G. TORSI, Equilibrium release of metal atoms at the electrothermal rod atomizer, *Annali di Chimica, 68,* pp. 943-966, (1978).

[2] S.L. PAVERI-FONTANA and R. RIGACCI, A singular perturbation weakly-singular integro differential problem from analytical chemistry, in: P.W. HEMKER and J.J.H. MILLER, *Numerical analysis of singular perturbation problems,* Acad. Press, 1979, pp. 475-484.

[3] S.L. PAVERI-FONTANA and G. TESSARI, Models in electrothermal atomization, the platform atomizer, *Progr. Analytical Atomic Spectroscopy, 7,* pp. 243-273, (1984).

[4] S.L. PAVERI-FONTANA, A. SRIVASTAVA and G. TESSARI, Models in electrothermal atomization: the open atomizer, in prepartion.

[5] S.L. PAVERI-FONTANA, Travelling waves for the enclosed atomizer, *Proceedings of the V-th International Conference on Mathematical Modelling,* in press.

[6] C.S. BLACK, M.R. RIDDLE and J.A. MALCOMBE, A Monte-Carlo simulation for graphite furnace atomization of copper, *Applied Spectroscopy,* in press.

For more information contact S.L. Paveri-Fontana, Univ. of Rome, Italy.

VIBRATION OF ELECTRICAL CABLES

This is a small research project with a company producing electrical cables. The aim of the project is to analyze the dynamic behaviour of air suspended cables for example under the action of wind effects.

The mathematical model can be given by means of a system of nonlinear coupled differential equations. The dynamic behaviour can be mastered if the initial position and the initial velocity is given. The role of the mathematics is to derive the right mathematical formulation from physical principles. It also turns out that some serious numerical analysis is needed in order to be able to solve realistic problems. It is possible to simulate for example dancing phenomena of the cable and to locate the points where the tension of the cable grows critically. The results can be utilized in product development.

For more information contact Prof. J. Saranen, Oula University, Section of Mathematics, Faculty of Technology, Linnanmaa 90570 Oula, Finland.

PENETRATION OF SOLVENTS INTO GLASSY POLYMERS

The main feature of this phenomena is the existence of sharp front moving into the polymer and separating a virtually solvent-free region from a swollen penetrated region. Typically, this is a free boundary problem. The solvent transport mechanism in the swollen region can be modelled in different ways (see [1] for a survey). In the first mathematical model, proposed in 1978 [2], a simple Fick's Law was assumed for the solvent concentration, while the penetration rate (i.e. the velocity of the front) was related to the unknown value of concentration on the front via a power law. In a one-dimensional geometry such an equation can be generalized to $ds/dt = f[c(s(t),t) - q]$, where $x = s(t)$ is the location of the front, c denotes concentration, q is a threshold value for penetration. The function f is increasing and such that $f(0) = 0$. Mass balance along $x = s(t)$, i.e. $D\partial c/\partial x = c\, ds/dt$, $D =$ diffusivity, gives the second free boundary condition. Such a problem has been studied extensively in [3], assuming a constant value for c at the outer surface of the polymer. For references on the subsequent mathematical literature and for further generalization, see [4].

[1] G. CAMERA-RODA, G.C. SARTI Non-Fickian mass transport through polymers: a viscoelastic theory. To appear on Proc. of IX Int. Conf. on Transport Theory (Nelson ed.).

[2] G. ASTARITA, G.C. SARTI, A class of mathematical models for sorption of swelling solvents inglassy polymers, Polym. Eng. Sci. *(1978), pp. 388-395.*

[3] A. FASANO, G.H. MEYER, M. PRIMICERIO, On a problem in polymer industry: theoretical and numerical investigation, To appear on SIAM J. Appl. Math.

[4] A. FASANO, R. RICCI Penetration of solvents into glassy polymers, In: *Free Boundary Problems: Theory and Applications, (Bossavit, Damlamian, Frémond eds.), Res Notes Math., Pitman.*

For more information on chemical aspects constant prof. Giulio Cesare Sarti, Instituto di Impianti Chimici, Facoltà di Ingegneria, Università di Bologna, Viale Risorgimento 2, 40136 Bolgna (Italy).

SIMULATION OF STREET-TRAFFIC

Problem. The theoretical analysis of traffic within city limits is because of the complexity of the problem extremely difficult. Only special simple situations are treated till now. Also experimental research is not always possible. The advantage of simulation is, that the passage of every single car and the traffic conditions at every point (for example at every crossing) can be observed. Corresponding observations in reality are only possible with extreme high expense. Moreover special situations of the traffic flow can be repeated arbitrary often in contrast to reality. Also the effect of planned changes (for example, new streets, new one-way regulations, new traffic lights) are to be determined by simulation.

Mathematics. An essential restriction on simulation is the capacity of the storage of the computer and the needed CPU-time. The street-traffic has to be modeled so that both quantities are as small as possible and that valid statements on traffic flow can be made. Simulation models are formulated both in GPSS and in FORTRAN. The advantage of the simulation program in FORTRAN is, that the information on the net of streets and the traffic flow is an input for the program. Using GPSS the program has to be rewritten when simulating a new town. For the input to the FORTRAN-program forms are designed. A form "net of streets" containing for example the maximal number are designed. A form "net of the streets", traffic lights, a form "streets" containing the running-time and a code of the next crossing for every street, a form "street crossing" containing informations on traffic lights and priorities and a form "traffic lights". The form "traffic" flow contains informations on the starting point and a binary code of the route of a passage. Output can be obtained for every trace of a street crossing and every street, for example the mean number of cars waiting, the mean waiting time or corresponding frequency tables. The organisation of the FORTRAN program is similar to the internal organisation in GPSS. The FORTRAN program contains beside subroutines for setting initial values and statistical evaluations, six subroutines for setting initial values and statistical evaluations, six subroutines for entering and leaving the street net, a special street and a crossing.

Applications:

The simulation model was used to simulate the traffic flow in Kapfenberg, a streetnet with 14 crossings, needing 13KBytes for the street net and 10KBytes for 500 cars. A traffic planner in Innsbruck uses the FORTRAN program since 1980.

[1] R. TAKACS, Ein Simulationsmodell für den Straßenverkehr, Doctoral Theses, Linz 1980.

For more information contact R. Takacs, Institute of Mathematics, University of Linz, Austria.

A PARAMETER ESTIMATION AND SIMULATION PROBLEM IN CHEMICAL KINETICS

We have the following problem.

We are interested in finding out the kinetic parameters of the following reaction scheme

$$A + B \xrightarrow{k_1} C + \Delta H_1$$

$$B + C \xrightarrow{k_2} D + \Delta H_2$$

i.e. we have four components and two reactions which are both exothermic releasing energies ΔH_1 and ΔH_2. The component C is the desired product and the final aim is to be able to simulate the reaction in different tempatures and with different ways of feeding A or B into the reactor in order to find out if there would be any cooling problems in the large scale production of chemical C.

An extra problem is caused by the fact that the components from A to D could not be analyzed and only the temperture in the reactor and of the cooling water could be measured. In addition to that ΔH_1 was approximately known and the kinetic parameters were assumed to obey the Arrhenius's law.

On the basis of the above knowledge a set of experiments with different initial concentrations of A and B were conducted in a reactor system where the temperature in the reactor and the temperature of the cooling water could be measured.

In its general form the problem is following: we have a model

$$\frac{ds}{dt} = h(t,s,z,\theta) \tag{1}$$

and observations

$$y_i = g(s(t_i,z_i,\theta),t_i,z_i,\theta) + \epsilon_i, \qquad i,\ldots,n \tag{2}$$

where (1) gives the differential equations that define the model. In the specific case discussed in this article these are nonlinear and time invariant. In this case also the observation function g depends on t and z only through s. ϵ_i is the error vector that is assumed to be normally distributed with an unknown covariance matrix V. θ is the unknown parameter vector to be estimated on the basis of the basis of the observations $y_i, i=1,\ldots,n$. The kinetic model for the special case described above is

$$\frac{ds}{dt} = h(s,x,\theta)$$

$$s(0) = s_0$$

$$y_i = As(t_i,x_i,\theta) + \epsilon_i, \qquad i=1,\ldots,n$$

$$A = \begin{bmatrix} 0\,0\,0\,0\,1\,0 \\ 0\,0\,0\,0\,0\,1 \end{bmatrix}$$

$$\epsilon_i \sim N(0,V)$$

where

$$s_1 = A,\ s_2 = B,\ s_3 = C,\ s_4 = D,\ s_5 = T,\ s_6 = T_w$$

T is the temperture in the reactor and T_w the temperture of the cooling water i.e. the measurements.

The form of h will not be specified here. The parameter vector θ contains the parameters of the Arrhenius's equation and in addition some other unknowns e.g. heat conductivities.

If we consider equations (1) in their solved form (in practise the solution is usually only by numerical methods) we can combine equations (1) and (2) to get a more familiar form of the problem, namely, let

$$f(x_i,\theta) = g(s(t_i,z_i,\theta),t_i,z_i,\theta)$$
$$x = [t_i,z']'$$

With this notation combining (1) and (2) gives

$$y_i = f(x_i,\theta) + \epsilon_i, i = 1,...,n \quad (3)$$

which is often called the standard reduced form of the nonlinear regression problem.

The estimates for the parameters were found by maximum likelihood estimation assuming the errors to be independently and normally distributed. After having found the estimates for the kinetic parameters simulations with the large scale system model were carried out. These simulations showed that the large scale production was possible with conventional cooling equipment. In addition some simulations modeling hazardous situations like cooling system break down were conducted in order to estimate the safety requirements. The major benefit of the modeling effort was that we did not have to conduct expensive pilot experiments.

For more information contact Veli-Matti Taavitsainen, Espoo Research Center, P.O. BOX 44 02271 Espoo, Finland.

THE IMPREGNATION OF PAPER BY A RESIN

Description In manufacturing laminates it is required to uniformly impregnate sheets of absorbent paper with a given amount of resin per unit area. The sheets are then presses together to form the final material. One simple procedure is to feed the paper into the bottom of a tank of the resin and use the effects of capillarity and hydrostatic pressure to achieve impregnation. This can be significantly improved, and greater control obtained, by feeding the paper through a pair of rollers rotating in opposite senses and using the resin as a fluid lubricant between the rollers and the paper. If the gap between the rollers is small high pressures will be developed so that in addition to impregnating large quantities of resin the paper will be compressed. As such pressures, for typical resins and papers, the rate of impregnation will be kp and the amount of compression cp where k and c are porosity and compressibility coefficients which in some sense are small. Optimum operating conditions occur when the uncompressed paper is thicker than the minimum gap between the rollers but is crucially important that the fluid layer should lubricate the whole section under the roller since the paper tears on contact with the roller surface.

Model It is appropriate to use classical lubrication theory for a viscous fluid as a model with additional terms representing the compression of the paper and the loss of fluid into the paper. A modified Reynolds equation may be obtained in the form

$$\frac{d}{dx}(h - \frac{h^3}{12}\frac{dp}{dx}) = -kp$$

where $h = 1 - a_0 + \frac{x^2}{2} + cp$ is the fluid film thickness and so the ratio of the paper thickness to the minimum gap between the rollers. Suitable boundary conditions are $p(-\infty)=0, p(d)=0=\frac{dp}{dx}(d)$, where the latter condition at an unknown point $x=d$ implies the existence of a cavity in $x>d$.

Analysis. By an appropriate scaling the equation may be reduced to

$$\frac{1}{12}\frac{d}{dx}[(sgn(1-s_0) + \frac{x^2}{z} + \bar{c}p)^3\frac{dp}{dx}] = \bar{k}p + x + \bar{c}\frac{dp}{dx}$$

and with $1-s_0>0$ a regular expansion in powers of $\bar{c}$ and $\bar{k}$ leads to impregnation rates which are too small. However with $1-s_0<0$ the coefficient of the highest derivative will vanish if the small parameter $\bar{c}$ is put equal to zero so that it is a singular perturbation problem. Four different regions are needed to describe the solution $-\infty<x,<\sqrt{2}, x\sim-\sqrt{2}+0(\epsilon^{\frac{1}{2}}), -\sqrt{2}<+\sqrt{2}$ and $x\sim-\sqrt{2}+0(\epsilon^{\frac{1}{2}})$; and in $-\sqrt{2}<x<,+\sqrt{2}$ and in $-\sqrt{2}<x<+\sqrt{2}$ the pressure is given by $\frac{1}{c}(1-x^2/2)$ so that is a simple matter to calculate the rate of impregnation and fluid film width. The solution may be obtained in each of these regions correct to $0(1)$ and matched asymptotic expansions. the condition that both the compressibility and porosity parameters are significant is that $k\sim c^{3/2}$ and if the fluid channel is not to close it is necessary that $k<\frac{c^{3/2}}{2^{3/4}(s_0-1)^{7/4}}$. From the detailed scalings it is a simple matter to determine how the impregnated mass varies with the angular velocity of the rollers and the minimum gap between them.

For more information contact A.B. Tayler, Oxford Study Group with Industry, Math. Inst., Univ. of Oxford, 24-29 St Giles, Oxford England.

THE DYNAMICS OF A CURRENT COLLECTION SYSTEM FOR AN ELECTRIC LOCOMOTIVE

Description. An electric locomotive coolects current from an overhead wire through a mechanical device called a pantograph. This consists of stiff springs and dampers in the shape of a rhombus which is mounted symmetrically on the locomotive roof with the top corner of the pantograph in contact with the hanging cable. The cable may be suspended in a numer of ways, the simplest of which is through a stiff ring onto equi-spaced overhead supports which may be considered to be fixed. The pantograph head imparts a moving disturbance to the hanging wire which depends on the force between the pantograph head and the wire through the pantograph dynamics. The problem is to determine the motion of the wire and to ensure that contact can be maintained between the pantograph head and the wire.

The most practical case is for a train moving with a constant speed which is less than the wave speed of small transverse oscillators in the wire.

However, a further interesting problem is that of a rocket suspended in a cradle hanging from an overhead wire, which accelerates from rest through the wave speed of small transverse oscillations in the wire. In this situation it is observed that the wire is likely to break into two pieces.

Model. It can be shown that away from the cable supports the non-dimensional wire displacement $y(x,t)$ from the static hanging position satisfies

$$\frac{1}{c^2}\frac{\partial^2 y}{\partial t^2} - \frac{\partial^2 y}{\partial x^2} = P(t)\delta(x-t), \quad 0<x<1$$

with $y(0,t) = 0 = y(1,t)$. The contact condition with the pantograph is

$$y(t,t) = Y(t) - 4t(1-t),$$

where $Y(t)$ is the displacement of the pantograph head. From the pantograph dynamics in a typical case

$$P = P_0 - b\frac{dY}{dt},$$

but other pantograph designs may give different relationships between P and the derivatives of Y.

Analysis. Careful use of simple solutions of the wave equation in various domains of the x,t plane leads to a solution for the wire displacement which is piecewise linear in a sequence of time intervals subdividing the time taken for the train to traverse one span. This sequence arises because of the multiple reflections between the pantograph head and cable support which it is approaching, and is an infinite sequence which leads to a divergent displacement unless

$$b < \frac{1}{2}(1+\frac{1}{c})^2.$$

The solution is periodic in each span but the vertical pantograph head velocity $\frac{dy}{dt}$ is discontinuous on crossing the support at $x = 1$. A more detailed analysis is therefore required in the vicinity of a support which requires a more complicated model for the pantograph dynamics and cable support condition. This model leads to an extremely interesting functional differential equation in the form

$$\frac{d\underset{\sim}{u}}{dt} = A\underset{\sim}{u}(t) + B\underset{\sim}{u}(\mu t),$$

whose properties are somewhat unexpected. Numerical timestepping methods may however be used to give good results for the consequent pulse profiles which are propagated along the wire. In the case of an accelerating moving source for which

$$\frac{dP}{dt} = -a - b\frac{d^2 Y}{dt^2},$$

the model does imply that a discontinuity in Y is propagated from the source when it attains the wire wave speed.

For more information contact A.B. Tayler, Oxford Study Group with Industry, Math. Inst., Univ. of Oxford, 24-29 St Giles, Oxford England.

RESONANCE IN A GEARBOX

Description. The primary functions of a gearbox is to effect a change in angular speed and to transmit power from the driving mechanism to the mechanism being driven or output device. This may be achieved by a sequence of gearwheels and pinions which mesh into each other through teeth on their perimeter. The gear ratio of a gearwheel and pinion measures the number of teeth on the gearwheel for each tooth on the pinion, and with large numbers of teeth and circular cross-sections this is equal to the ratio of the radii of the gearwheel to that of the pinion. The gearwheel may wear after prolonged use and lose its circular shape giving rise to a resonance oscillation which can create large enough torques to strip the gears. The problem is to determine these resonant frequencies and to estimate the torque between the gearwheel and pinion in a state of resonance in terms of the amount of wear. It is then possibly by measuring the wear to determine when to replace or recut the gearwheel before it is stripped of its teeth.

Model. For a single perfect gearwheel and pinion their angular positions θ_3 and θ_2 are connected by $\theta_2+\gamma\theta_3=0$ where γ is the gear ratio. If the driving mechanism has an angular position θ_1 and is connected to the pinion by a shaft of stiffness k_1, then

$$\frac{I_1 d^2\theta_1}{dt^2}=k_1(\theta_2-\theta_1) \text{ and } \frac{I_2 d^2\theta_2}{dt^2}=k_1(\theta_1-\theta_2)+P,$$

where P is the force between the pinion and gearwheel divided by the pinion radius. If the output has angular position θ_4 and is connected to the gearwheel by a shaft of stiffness k_2 then also

$$\frac{I_3 d^2\theta_3}{dt^2}=k_2(\theta_4-\theta_3)+P, \quad \frac{I_4 d^2\theta_4}{dt^2}=k_2(\theta_3-\theta_4).$$

Hence $\gamma(I_1\theta_1+I_2\theta_2)-I_3\theta_3-I_4\theta_4=ct$, and a steady state motion exists with

$$\theta_1=\theta_2=-\gamma\Omega t, \quad \theta_3=\theta_4=\Omega t.$$

For a worn gearwheel $\theta_2+\gamma\theta_3=\epsilon f(\theta_3)$ where $\epsilon<<1$ and $f(\theta_3)$ is a periodic function, period 2π. In general this small wear term will imply a small correction to the force P unless it stimulates a possible free oscillation of the unperturbed system when a much larger force will result due to the resonant interaction.

Analysis. Free oscillations of the form $\theta_j=a_j e^{iwt}$ may be superposed on the steady state motion if w is a root of the determinant

$$\begin{vmatrix} k_1-w^2I_1 & -k_1 & 0 & 0 \\ 0 & 1 & \gamma & 0 \\ 0 & 0 & -k_2 & k_2-w^2I_4 \\ \gamma I_1 & \gamma I_2 & -I_3 & -I_4 \end{vmatrix}=0$$

and two such roots for w^2 do exist. If we choose to examine the interaction of the nonlinear term $\epsilon f(\theta_3)$ with one of these free oscillations, say w_1, then we consider $\Omega=w_1+\delta\epsilon$ and focus attention on the amplitude of the eigenvector corresponding to the eigenvalue $w^2=w_1^2$. This amplitude satisfies the equation

$$\frac{d^2u}{dt^2}-w_1^2u=\lambda\epsilon\sin((\omega_1+\epsilon\delta)t+u)$$

where γ is a known constant and the simplest wear function which is periodic, period 2π, has been used.

A regular perturbation is powers of ϵ clearly leads to a secular term at the $0(\epsilon)$ stage so that such an expansion will not valid for times of $0(1/\epsilon)$. The method of two-timing, in which a second time $\tau=\epsilon t$ is introduced, gives a uniformly valid solution for all t in the form

$$u \sim a(\tau)\sin w_1 t + b(\tau)\cos w_1 t + 0(\epsilon),$$

where a and b satisfy a pair of first order differential equations in τ. A phase plane analysis then gives bounds on $a^2 + b^2$ for all initial data at $\tau = 0$, and hence on P.

For more information contact A.B. Tayler, Oxford Study Group with Industry, Math. Inst., Univ. of Oxford, 24-29 St Giles, Oxford England.

ELECTROCHEMICAL PAINTING

Description. A convenient method for painting metal objects is by immersing the workpiece in an electrolyte solution containing large negatively-charged organic ions. The ions are neutralized by imposing a potential difference between the workpiece as a node and the earthed bath containing the electrolyte. The applied voltage induces a number of electrochemical and transport processes which result in all or part of the workpiece being painted. However, not all the immersed surfaces of the workpiece are necessarily painted and if the voltage is switched off the paint deposited on the surface will dissolve back into the electrolyte solution. Two separate phases are observed in the deposition process, in the first of which rapid changes take place within the bath and painting starts. In the second phase, which is much slower, the paint layer is built up and appears to approach an almost steady state, that is regions which are unpainted in minutes have changed only a little in hours. The problem is to determine whether all the surface is painted or not after a reasonable process time and to obtain some estimate of the thickness of paint film which has been deposited. A further problem is to consider modifications to the geometry of the workpiece and the bath to improve the painting characteristics.

Model. The process is complicated but we focus on the second phase and look for steady state conditions. The concentration of paint ions in the bath is assumed to be constant due to stirring and gross change neutrality is maintained so that the electric potential satisfies $\nabla^2\phi=0$ where the electric current is $\underline{j}=-\sigma\nabla\phi$. This is also assumed to hold within the paint layer adjacent to the workpiece where the conductivity $\bar{\sigma}$ will be much smaller than σ, possibly $\bar{\sigma}/\sigma=10^{-6}$. Hence the potential gradient across the paint layer will be large and approximately constant, and the thickness of the paint layer h small. If ϕ_s is the potential difference across the paint layer then

$$h \approx \frac{\bar{\sigma}\phi_s}{j_n} = \frac{\bar{\sigma}}{\sigma}\frac{\phi_s}{(\frac{\partial\phi}{\partial n})}.$$

Since with no current flowing there is a dissolution of paint at a rate d assumed constant there must be a dissolution current j_0 which would exactly balance the outflow of negative ions. Thus in a steady state the boundary condition at the surface of the painted layer, which is effectively the surface of the workpiece, is $\frac{\partial\phi}{\partial n}=j_0$, where j_0 is to be evaluated from experimental data. Where the workpiece is unpainted it must have constant potential, and we choose $\phi=0$ at the workpiece and 1 on the bath. This leads to the boundary value problem

$$\nabla^2\phi = 0 \text{ in } D, \quad \phi = 1 \text{ on the bath;}$$

$$\phi = 0, \ \frac{\partial\phi}{\partial n} < j_0 \text{ on the unpainted workpiece;}$$

$$\phi > 0, \ \frac{\partial\phi}{\partial n} = j_0 \text{ on the painted workpiece.}$$

Analysis. This may be rephrased as a variational inequality in the form

$$\int_D \nabla\phi\cdot\nabla(v-\phi)d\tau + j_0\int(v-\phi)dS \geqslant 0$$

for all admissible functions v such that $v = 1$ on the bath, $v \geqslant 0$ on the workpiece, and suitable continuity properties. If in addition v belongs to a convex set this inequality has a unique solution which determines the regions on the workpiece which are painted and unpainted. It may also be rephrased as finding the extremal $v = \phi$ which minimises

$$E(v) = \frac{1}{2} \int_D |\nabla v|^2 d\tau + j_0 \int_{\text{workpiece}} v dS,$$

and gives a convenient numerical procedure for the solution.

For more information contact A.B. Tayler, Oxford Study Group with Industry, Math. Inst., Univ. of Oxford, 24-29 St Giles, Oxford England.

THE WELDING OF TWO MILD STEEL PLATES

Description. In many manufacturing processes mild steel plates have to be welded together with great precision and a considerable number of welds may be needed as for example in making a car body. One convenient way of making an individual weld is to apply electrodes to either side of the two plates pressed together and pass a current. The current will provide a volumetric heat source within the thickness of the plates and after sufficient time has elapsed melting will occur at the interface between the two plates where by symmetry the temperature is a maximum. Thus a molten nugget will form and will grow outwards towards the cooled outer faces. The problem is to determine this growth so that the current can be switched off when the nugget is large enough to form a sufficiently strong weld but not so large that there is a risk of molten metal coming into contact with the electrodes. The welding procedure is likely to be automated and it is important to establish the welding time in terms of the material properties and the current applied. Usually an alternating current is used and the electrodes have a large constant inductance so that the current amplitude is essentially constant. The timescale of the welding process is much longer than the a.c. cycle time and in the first phase of the process below about 400°C there are a number of complicated effects including variable resistivity, air resistance between the surfaces, and current transients. However, after this starting phase when temperatures are above 400°C these effects do not appear to be important and all the material properties are effectively constant.

Model. The problem is fully three dimensional although possibly axisymmetrical but we consider a one dimensional model which will give an upper bound for the nugget growth and a reasonable estimate for its thickness on the electrode centre line (given that the plates are thin). The nondimensional temperature $u(x,t)$ satisfies

$$\frac{\partial u}{\partial t} = \frac{\partial^2 u}{\partial x^2} + q, \quad 0<x<1,$$

where q is the constant heat source term due to the current. By symmetry $\frac{\partial u}{\partial x}=0$ at $x=0$ the interface between the plates, and at $x=1$, the interface between the plate the electrode,

$$-\frac{\partial u}{\partial x} = H(u-f),$$

where f is the known mean temperature of the electrode and H is a heat transfer coefficient obtained empirically. Initially (in this second phase) we assume that the temperature is known so that

$$t=0, \quad u=g(x)<0,$$

where u has been measured relative to the melting temperature. To complete the problem the melting phenomenon must be modelled and if there is a sharp interface between solid and liquid states the Stefan condition

$$u=0, \ \left[\frac{\partial u}{\partial x}\right]_{x=s-}^{x=s+} = \lambda\frac{ds}{dt}, \quad \text{on } x=s(t),$$

must apply where λ is a non-dimensional latent heat. In this problem however such a formulation may be shown to be ill-posed and a mushy region results in which the temperature $u \equiv 0$ and the material composition is part solid and part liquid.

Analysis. To avoid the difficulty of specifying continuity conditions on the boundaries of a mushy region we rephrase the problem in weak solution form. Thus we look for a pair of functions u and h satisfying $h=u$, $u<0$, $h=u+\lambda$, $u>0$; and

$$\int_0^{t_0}\int_0^{t}\left(h\frac{\partial\psi}{\partial t} + u\frac{\partial^2\psi}{\partial x^2} + q\psi\right)dx\,dt + \int_0^{1}\psi_{t=0}\,g\,dx + \int_0^{t_0} f\psi_{x=1}\,dt = 0$$

for all admissable functions v satisfying $\frac{\partial v}{\partial x}=0$, $x=0$; $\frac{\partial v}{\partial x}+Hv=0$, $x=1$; $v=0$ on $t=t_0$; together with appropriate continuity conditions. It may be shown that this reduces to the boundary value problem where u is twice differentiable and satisfies the Stefan condition across lines of discontinuity in $\partial u / \partial x$. Moreover it has a unique solution which may be obtained by an iterative procedure, now called the enthalpy method, in which at each step in the iteration u is obtained from a knowledge of h thus avoiding any non-uniqueness.

For more information contact A.B. Tayler, Oxford Study Group with Industry, Math. Inst., Univ. of Oxford, 24-29 St Giles, Oxford England.

MULTI PLANE BALANCING PROBLEM

Balancing of large rotating machines is frequently possible to carry out without removing the rotor. The unbalance can be detected indirectly with vibration measurements that will produce data for compensating the inbalance. The inbalance is strictly speaking dynamic and the rotor of the rotating machine is elastic. Hence the general solution of the balancing problem depends on the rotating frequency and the critical frequencies of the machine.

The Central Laboratory of Imatran Voima Oy has developed a multi plane balancing method based on vibration measurements. The mathematical algorithm has been written in FORTRAN-77. This balancing method has been applied successfully in reducing harmful vibrations of turbogenerators.

For the balancing procedure one must select the vibration measurement planes and balancing planes onto which discrete masses will be inserted. In order to use the so-called action coefficient method one determines the action of a trial mass inserted to each balancing plane in turn on each measurement quantity. Assuming the action to be linear in masses the balancing problem can be solved by minimizing the quadratic sum of inbalances in the balancing masses.

The function to be minimized is

$$M = -[\underline{S}^+\underline{S}]^{-1}\underline{S}^+\underline{\mathrm{v}}_0,$$

where

$$\underline{S} = [S_{kl}] = \frac{1}{m_{l0}e^{j\phi_{l0}}}[\mathrm{v}_{kl}e^{j\phi_{kl}} - \mathrm{v}_{k0}e^{j\phi_{k0}}],$$

$$\underline{\mathrm{v}}_l = [\mathrm{v}_{1l}e^{j\phi_{1l}},\ldots,\mathrm{v}_{kl}e^{j\phi_{kl}},\ldots,\mathrm{v}_{Nl}e^{j\phi_{Nl}}]^T.$$

S is the action coefficient, $k = 1,\ldots,N$ (the number of measurement planes), $l = 1,\ldots,M$ (the number of balancing planes), v_{kl} is the amplitude and ϕ_{kl} is the phase of vibration in measurement plane k induced by the trial mass

$$m_{l0}e^{j\phi_{l0}}$$

inserted onto the balancing plane l. The term v_0 refers to the original inbalance. Practically one must set restrictions on the balancing masses and remaining inbalances which are both complex variables. Weighting factors for measurement planes are necessary in order to compensate effectively the most serious inbalances. In the general problem setting one must therefore optimize a quadratic sum in a subset of a multidimensional complex space. The terms of the sum are restricted in magnitude and have weighting factors selected according to importance of each inbalance. Furthermore the overall process makes the vibration measurements noisy so that the statistical nature must be taken into account.

For more information contact Tuomo Tilli, Imatran Voima Oy, IVO Central Laboratory, P.O. Box 112, 01601 Vantaa, Finland

THE COMPUTATION OF THE HYDRODYNAMIC COEFFICIENTS FOR PROBLEMS WITH ROTATIONAL SYMMETRY

Problem. The motion of a solid in an inviscid, incompressible and irrotational fluid is considered. For a general motion of the solid the effect of the presence of the fluid results in a virtual increase of the kinetic energy of the solid, represented by three second-rank tensors whose components are called the hydrodynamic coefficients. For problems with rotational symmetry four distinct non-zero hydrodynamic coefficients are to be computed. In applications, these coefficients are used, e.g., for studying critical vibrations of turbines.

Mathematics. Eventually the above problem leads to the solution of three Neumann boundary value problems for Laplace's equation. These problems are solved by a finite element method with rotationally symmetric elements. The four hydrodynamic coefficients are certain easily computable functionals of the solution of the Neumann problems.

[1] W. ZULEHNER, Über die Berechnung der hydrodynamischen Koeffizienten für den rotationssymmetrischen Fall, Diploma thesis, Dept. of Mathematics, University of Linz, 1978.

For more information contact W. Zulehner, Working group on industrial and business mathematics, Universität Linz, Austria.

NONLINEAR IDENTIFICATION OF INDUSTRIAL PROCESSES

Nonlinear modelling and identification are of great importance in many engineering designs to represent the dynamical behaviour at certain industrial processes for important variations of working points and disturbances.

Modelling by linear identification is only valid around a working point and cannot represent dynamical behaviour at a large scale of practical working conditions.

The construction of a nonlinear model using physical equations and its adjustment by experimental measures is often not easy and all this, requires considerable computing time.

The nonlinear identification problem is far from being satisfactorily solved in spite of all the techniques that have been developed over the last two decades. Amongst all these various techniques, none can be recommended, the choice being strongly dependent on the physical process to be identified. For example, the one using physical equations is often not easy since the number of parameters to be adjusted by means of experiments rapidly grows with the complexity of the process.

The proposed technique uses extensions with polynomial inputs of bilinear systems which are called state-affine systems; it is applied to processes that have a linear representation around any operating point. The main interest of nonlinear modelling is to be able with one global model, to follow the dynamic and static evolution of the process over a wide range of operating conditions for eventually large variations of the controls; this objective is achieved.

Various examples dealing with electrical hydraulic, thermal and even nuclear power plants have been shown (Sections 1, 2 and 3).

This method is actually applied to the helicopter flight (Section 4).

1. Modelisation of the wave propagation in an hydraulic power plant:

This concerns a single in- and output processes. The control is the water flow rate turbined by the electric plant. The output is the amplitude of the wave at the level to control. We obtain a state-affine model defined on R^3 which remains valid for different values of the main flow rate (operating point) and enables us to represent oscillating responses as well as damped responses. [1]

2. Modelisation of the superheater temperature in a thermal power plant:

The goal is to identify the variations of the superheater steam temperature by acting on the water spray rate for different powers of the plant (working point). Two types of superheaters have been studied for which we obtain state-affine models defined on R^3 and R^4. In order to test the validity of these models for transitory operating points not used in the identification procedure, we have simulated a PRBS on the control and simultaneously a linear variatioan of the power of the plant. [1]

3. Modelisation of the steam exchanger in a nuclear power plant (first French breeder):

This concerns several in- and output processes. The controls are the steam valve aperture, the water flow rate and the sodium flow rate. The outputs are the steam pressure, the sodium temperature and the power of the turbine. In this case, the operating mode is the main steam flow rate. Two single in- and output state affine models defined on R^5 are obtained for representing the actions of sodium flow rate on the steam pressure and the sodium temperature. In [2] and [3], this method is generalized to the multivariable situation.

In this case, it is necessary to obtain the global form representation by using algorithm reduction of state space size.

This model remains valid for large steps and PRBS control parallel to important variations of the load and moreover this model is very effective to represent asymetry of the control.

4. Modelisation of the helicopter flight

In this application, the helicopter is assumed to be in an equilibrium position (constant altitude and constant speed with respect to the ground). We study the effect of the variations of the main pitch on the first component of the speed with respect to the air. The scale of speed of the helicopter is from 100 to 300 km/h.

The first results enable us to think that the state-affine model will be defined on R^3 or R^4.

[1] DANG VAN MIEN H. and D. NORMAND-CYROT (1984). Nonlinear State-Affine Identification Methods: Applications to Electrical Power Plants, Automatica, vol.20, No 2, pp. 175-188, 1984.

[2] BOURDON P., H. DANG VAN MIEN AND D. NORMAN-CYROT (1983). Applications to Nuclear Power Plants of Nonlinear Identification and Realization Techniques, VI IFAC Symp. on Identification and System Parameter Estimation, Washington, 2, 1514-1519.

[3] BOURDON P., H. DANG VAN MIEN and D. NORMAN-CYROT (1983). Nonlinear Modelling of Electric Power Plants, Proceedings of the IFAC Workshop, Como, Italy, 22-23 September 1983.

List of Participants

S.T.M. Ackermans
Eindhoven Univ. of Techn.
Sansovinostraat 1
5624 JW Eindhoven
The Netherlands

G. Alberts
Eindhoven Univ. of Techn.
Postbus 513
5600 MB Eindhoven
The Netherlands

J.G. Andrews
Marchwood Engineering Labs.
CEGB, Marchwood
Southampton, SO4 4ZB
United Kingdom

C. Bardos
Ecole Normale Superieure
45 rue d'Ulm
75009 Paris
France

J.F. Benders
Eindhoven Univ. of Techn.
Postbus 513
5600 MB Eindhoven
The Netherlands

A. Bensoussan
I.N.R.I.A.
Domaine de Voluceau de Rocquencourt
B.P. 15
78150 Le Chesnay
France

M.I.G. Bloor
Dep. of Appl. Math.
University of Leeds
Leeds, LS2 9JT
United Kingdom

J. Boersma
Eindhoven Univ. of Techn.
Postbus 513
5600 MB Eindhoven
The Netherlands

J.A.M. Bollen
Vredestijn Banden B.V.
Ir. Schiffstraat 370
7547 RD Enschede
The Netherlands

G. Bornard
LAG Grenoble
B.P. 46
38402 St. Martin d'Heres
France

V. Capasso
Univ. of Bari (CSATA)
Str. Prov. per Casamassima KM3
70010 Valenzano (Bari)
Italy

E. Cumberbatch
Claremont Graduate School
Claremont, CA 91711
USA

T. Donegan
Dep. of Math.
Univ. of Ulster at Jordanstown
Shore Road
Newtonabbey, Co. Antrim, BT37 0QB
Ireland

F. Eising
Hollandse Signaalapparaten B.V.
Dep. SEAT
Postbus 42
7550 GD Hengelo
The Netherlands

Ph. van Elteren
Katholieke Univ. Nijmegen
MSA Toernooiveld 1
6525 ED Nijmegen
The Netherlands

J.J.M. Evers
Dep. Appl. Maths.
Twente Univ. of Techn.
Postbus 217
7500 AE Enschede
The Netherlands

A. Fasano
Institute of Maths.
Univ. of Florence
Viale Morgagni 67/a
50134 Florence
Italy

B.H. Gilding
Dep. of Appl. Maths.
Twente Univ. of Techn.
Postbus 217
7500 AE Enschede
The Netherlands

F.A. Goldsworthy
Dep. of Appl. Maths.
Univ. of Leeds
Leeds, LS2 9JT
United Kingdom

J. de Graaf
Eindhoven Univ. of Techn.
Postbus 513
5600 MS Eindhoven
The Netherlands

W.A. Green
Dep. of Theor. Mech.
Univ. of Nottingham
Nottingham, NG7 2RD
United Kingdom

P. ten Hagen
Centrum voor Wis. & Infor.
Kruislaan 413
1098 SJ Amsterdam
The Netherlands

A. van Harten
Inst. of Maths.
Rijksuniversiteit Utrecht
Budapestlaan 6
3584 CD Utrecht
The Netherlands

M. Heilio
Lappeeranta Univ. of Techn.
P.O. Box 20
6351 Lappeenranta 80
Finland

J.R. Heringa
Grevingaheerd 172
9737 ST Groningen
The Netherlands

F. Hodnett
Dep. of Maths.
National Inst. of Higher Educ.
Limerick, Ireland

R. van der Hout
AKZO Research, Afd. CRS
Postbus 60
6800 AB Arnhem
The Netherlands

S. Howison
Math. Inst.
Oxford Univ.
24 St. Giles
Oxford, OX1 3LB
United Kingdom

F.J. Jacobs
Koninklijke Shell
Expl. Prod. Lab.
Postbus 60
2280 AB Rijswijk
The Netherlands

I. Karpouzas
Inst. Biomedical des Cordeliers
(Medimat)
Univ. de Paris VI
15, rue de l'Ecole de Medicine
75270 Paris cedex 06
France

J. Koehl
MPI fur Mathematik
Gottfried-Claren-Strasse 26
D-5300 Bonn 3
FRG

G.J.G. Kok
Haskoning
Royal Dutch Consulting Engineers
P.O. Box 151
6500 AD Nijmegen
The Netherlands

J.P.M. de Kroon
Ned. Philips Bedrijven B.V.
Centre for Quantitative Methods
Geb. VN 703
Postbus 218
5600 MB Eindhoven
The Netherlands

A. Lahtinen
Dep. of Maths.
Univ. of Helsinki
Hallituskatu 15
00100 Helsinki 10
Finland

J. de Lange
OW & OC
Rijksuniv. Utrecht
Tiberdreef 4
3561 GG Utrecht
The Netherlands

C.B. Leeftink
Masterfoods C.V.
Roosstraat 100
333 SM Zwijdrecht
The Netherlands

J.K. Lenstra
Centrum voor Wisk. & Infor.
Kruislaan 413
1098 SJ Amsterdam
The Netherlands

P. van Loon
Eindhoven Univ. of Techn.
Postbus 513
5600 MB Eindhoven
The Netherlands

H.H. Martens
Norwegian Inst. of Techn.
7034 Trondheim
Norway

S. McKee
UCINA
Oxford Univ. Comp. Lab.
8-11 Keble Road
Oxford, OX1 3QD
United Kingdom

J. Molenaar
Wisk. Dienstverlening
Kath. Univ. Nijmegen
Toernooiveld
6525 ED Nijmegen
The Netherlands

H. Neunzert
Univ. of Kaiserslautern
675 Kaiserslautern
FRG

G.J. Olsder
Dep. of Maths. & Comp. Sc.
Delft Univ. of Techn.
2600 AJ Delft
The Netherlands

D.H. Owens
Dep. of Maths.
Univ. of Strathclyde
Richmond Street
Glasgow
United Kingdom

L.A. Peletier
Rijksuniv. Leiden
Wassenaarseweg 80
2333 AL Leiden
The Netherlands

J. Petersen
AUDI AG
Abt. I/EGA
Postfach 220
8070 Ingolstadt
FRG

R.M. Pidatella
Dep. of Maths.
Univ. of Catania
Viale A. Doria 6
95125 Catania
Italy

J. Piepers
Koninklijke Shell Lab.
Badhuisweg 3
1031 CM Amsterdam
The Netherlands

A. Plebe
Agriculture Indus. Develop.
Blocca Palma I Zona Industr.
Catania
Italy

J. Ponstein
Interfac. Econometrie
Rijksuniv. Groningen
Postbus 800
9700 AV Groningen
The Netherlands

S.W. Rienstra
Nat. Lucht- en Ruimtevaart Lab.
Voorsterweg 31
8316 PR Markenesse
The Netherlands

J. Roerdink
Centrum voor Wisk. & Infor.
Kruislaan 413
1098 SJ Amsterdam
The Netherlands

J. Rugelj
Inst. Jozef Stefan
Jamova 39
61000 Ljubljana
Yugoslavia

F.H. Ruymgaart
Maths. Inst.
Kath. Univ. Nijmegen
Toernooiveld
6525 ED Nijmegen
The Netherlands

L. Sant
Maths. & Sc. Dep.
Univ. of Malta
MSIDA
Malta

F. van Scheepen
European Commission
Dep. for Sc., Res. & Developm.
Wetstraat 200
Brussels
Belgium

W. Schempp
Lehrstul fur Math. I
Univ. of Siegen
D-5900 Siegen
FRG

J.H. van Schuppen
Centrum voor Wisk. & Infor.
Kruislaan 413
1098 SJ Amsterdam
The Netherlands

R. Semenzato
Dip. di Fisica "G. Galilei"
Univ. of Padova
Via Marzolo 8
35131 Padova
Italy

P. Siwak
Inst. Automatyki
Techn. Univ. of Poznan
Piotrowo 3a
60-965 Poznan
Poland

M. Skapin
Inst. Jozef Stefan
Jamova 39
6100 Ljubljana
Yugoslavia

A. van der Sluis
Rijksuniv. Utrecht
Domplein 29
3512 HD Utrecht
The Netherlands

J.H. de Smit
Dep. of Appl. Math.
Twente Univ. of Techn.
Postbus 217
7500 AE Enschede
The Netherlands

G. Soderlind
The Swedish Inst. of Appl. Math.
P.O. Box 26300
10041 Stockholm
Sweden

H.M. Soekkha
Center for Transport. Eng.
Delft Univ. of Techn.
Postbus 5048
2600 GA Delft
The Netherlands

M.J. Stanford
Nat. Phys. Lab.
Queens Road
Teddington, Middlesex TW11 0LW
United Kingdom

D. Sundstrom
The Swedish Inst. for Appl. Math.
P.O. Box 26300
10041 Stockholm
Sweden

N.M. Temme
Centrum voor Wisk. & Infor.
Kruislaan 413
1098 SJ Amsterdam
The Netherlands

C.R. Traas
Twente Univ. of Techn.
Postbus 217
7500 AE Enschede
The Netherlands

F. Verhulst
Dep. of Math.
Rijksuniv. Utrecht
Budapestlaan 6
3584 CD Utrecht
The Netherlands

W.K. Vietsch
Ministerie van O & W
Postbus 25000
2700 LZ Zoetermeer
The Netherlands

C.D. Vreugdenhill
Waterloopkundig Lab.
Postbus 152
8300 AD Emmeloord
The Netherlands

J.M.J. Waals
NV Waterinningsbedrijf
Brabantse Biesbosch
Postbus 61
4250 DB Werkendam
The Netherlands

Hj. Wacker
Math. Inst.
Univ. of Linz
Altenbergerstrasse
4040 Linz
Austria

T. Tomiyama
Centrum voor Wisk. & Infor.
Kruislaan 413
1098 SJ Amsterdam
The Netherlands

Organizing Committee

E.W.C. van Groesen
Dep. of Appl. Math.
Twente Univ. of Techn.
Postbus 217
7500 AE Enschede
The Netherlands

M. Hazewinkel
Centrum voor Wisk. & Infor.
Kruislaan 413
1098 SJ Amsterdam
The Netherlands

R.M.M. Mattheij
Dep. of Math.
Katholieke Univ. Nijmegen
Toernooiveld
6525 ED Nijmegen

Members of the committee for developing industrial mathematics in the Netherlands

Prof.dr. J.F. Benders
Onderafdeling der Wisk. en Infor.
Techn. Univ. Eindhoven
Postbus 513

Prof.dr.ir. W. Eckhaus
Rijksuniv. Utrecht
Subfac. Wisk. en Infor.
Budapestlaan 6
3584 CD Utrecht

Prof.dr. M. Hazewinkel
Centrum voor Wisk. en Infor.
Postbus 4079
1009 AB Amsterdam

Prof.dr. G.J. Olsder
Onderafd. der Wisk. en Infor.
Techn. Univ. Delft
Julianalaan 132
2628 BL Delft

Prof.dr. F.H. Ruymgaart
Katholieke Univ.
Subfac. Wisk. en Infor.
Afd. Wiskudige Dienstverlening
Toernooiveld
6525 ED Nijmegen

Prof.dr.ir. P.J. Zandbergen
Onderafd. der Toegepaste Wisk.
Techn. Univ. Twente
Postbus 217
7500 AE Enschede

Dr.ir. H. Bosma
N.V. Philips
Natuurkundig Lab.
Gebouw WB-5
Postbus 218
5600 MD Eindhoven

Prof.dr. W.C.A. Kok
Verzekeringskamer
Postbus 9029
7300 Apeldoorn

Dr. R.M.M. Mattheij
Math. Instituut
Katholieke Univ. Nijmegen
Toernooiveld
6525 ED Nijmegen

Ir. J. Piepers
Koninklijke/Shell Lab.
Badhuisweg 3
1031 CM Amsterdam

Dr. J.P.M. de Kroon
Ned. Philips Bedrijven
Centrum voor Quantitatieve Methoden
Gebouw VN-7.03
Postbus 218
5600 MB Eindhoven

Dr. J. Telgen
Van Dien en Co., Afd. Organisatie
Churchill-laan 11
3527 Utrecht

INDEX

Proceedings of the Conference

Mathematics in Industry

October 24–28, 1983 Oberwolfach

Edited by Prof. Dr. H. NEUNZERT, University of Kaiserslautern, W.-Germany

1984. 287 pages. 16,2×23,5 cm. ISBN 3-519-02610-4. Paper DM 52,–

Contents

MIX
Papier aus verantwortungsvollen Quellen
Paper from responsible sources
FSC® C105338

If you have any concerns about our products,
you can contact us on
ProductSafety@springernature.com

In case Publisher is established outside the EU,
the EU authorized representative is:
Springer Nature Customer Service Center GmbH
Europaplatz 3, 69115 Heidelberg, Germany

Printed by Libri Plureos GmbH
in Hamburg, Germany